# The Art of Engineering Leadership

Die Zugangsinformationen zum eBook Inside finden Sie am Ende des Buchs.

Michael Jantzer · Godehard Nentwig ·
Christine Deininger · Thomas Michl

# The Art of Engineering Leadership

## Compelling Concepts and Successful Practice

 Springer

Michael Jantzer
Robert Bosch GmbH
Stuttgart, Germany

Godehard Nentwig
ENGENCE Engineering Excellence
Kusterdingen & Stuttgart, Germany

Christine Deininger
ENGENCE Engineering Excellence
Kusterdingen & Stuttgart, Germany

Thomas Michl
Robert Bosch GmbH
Stuttgart, Germany

ISBN 978-3-662-60386-4      ISBN 978-3-662-60384-0   (eBook)
https://doi.org/10.1007/978-3-662-60384-0

This Springer imprint is published by the registered company Springer-Verlag GmbH, DE part of Springer Nature.
The registered company address is: Heidelberger Platz 3, 14197 Berlin, Germany

# Acknowledgements

This book is the result of a long journey that the authors have taken together with many discussion partners.

In 2007, Robert Bosch GmbH launched a program to design and spread a development system in response to a quality crisis. The program is based on the belief that teams and managers can make good decisions only on the basis of understood cause effect relationships. Initially, the focus was on the application of a systematic approach, on model-based product development.

Five years later, a second focal point was placed on the education of engineering leaders.

During the conception of the content we realized that we did not have a common understanding of leadership. There were significant differences in understanding between colleagues from software development and those from the development of automotive supplier parts. In controversial discussions, the concepts of hierarchical leadership and servant leadership clashed. At the same time, the world around us changed. Agile work organizations found their way into the industry.

The result of the struggle was a new understanding of leadership as a role that can be systematically shaped and learned. It enables managers to change roles and systems faster and more effectively.

Many people were involved in the conception and development of the entire program. We would like to emphasize in particular: Thomas Bähren, Olga Bangert, Martin Brett, Dieter Dannhauer, Alain Deleonardo, Volkmar Denner, Hartmut Dykmann, Jürgen Gamweger, Volker Haas, Oliver Hörrmann, Günther Hohl, Martin Hurich, Oliver Jöbstl, Andreas Karl, Georg Ketteler, Andreas Kerst, Vadiraj Krishnamurthy, Stefan Link, Marcel Munoz, Marianna Niranjan, Jens Pawlak, Günther Plapp, Ulrich Schopf, Sylvester Schmidt, Holger Sobanski, Jürgen Spachmann, Karsten Strehl, Martin Thomas, Wolfgang Thuss, Xiaojiang Yang and Norbert Weller.

Most of the content and models in this book were developed and evaluated during discussions with over 4500 managers at all levels and functions worldwide.

Vanessa Erlmoser (design), Ruth Ewertowski (editing), Gert Hägele (editing), Sebastian Kanne (editing), Andreas Mayer (graphics), and Mark Plencner (translation) have contributed to the book.

We are very grateful to them. Without their help this book would not have been possible.

The proceeds from this book will be donated to Primavera—Hilfe für Kinder in Not e. V. (Help for Children in Need), which is supported by Bosch associates on a voluntary basis.

Stuttgart                                                              Michael Jantzer
June 2019                                                          Godehard Nentwig
                                                                  Christine Deininger
                                                                       Thomas Michl

# Contents

# Introduction

**1**

Engineers develop new products, new technologies and change the world we live in. In order to accomplish this, engineering leaders change organizations and develop their teams. Their roles and tasks are ever evolving. The roles and tasks are shaped not only by the technical problems they solve together with their team. They are also shaped by the business model of their industry, the culture of their organization and the personalities within the team. The foundation of the role of an engineering leader is therefore to be a facilitator of change. The key to success is the combination of leadership skills, a methodical approach to the tasks, a deep understanding of the underlying technologies and a solid set of management skills.

Since product developers usually see the success of their work a long time after the completion, the development work always takes place in the field of the unknown. Metric based management is of little value if there is a big time lag between performance (development) and earnings (market penetration). Therefore, product design is based on scientific approaches (e.g. understood physical cause effect relations), proven practices and individual beliefs. Developing those beliefs is part of the personal growth of an engineering leader. Over the past five years, we and the team of trainers at our Product Engineering Academy have worked on these questions with more than 4500 Bosch leaders worldwide. We discussed and improved leadership roles and approaches in more than 9000 workshops. The core of our training course is the development of an individual leadership model, which enables a transparent discussion on targets expectations and measures. We discuss beliefs individually, in small groups. In the form of coaching or consulting. Our managers appreciate exploring new ways to learn the beliefs of colleagues, employees and superiors. Verbalizing beliefs, making them transparent and creating a coherent picture is hard work. It takes courage and maturity to review this picture with fellow leaders, but the ensuing progress is usually stunning. In our opinion, this is a central contribution on the way to effective leadership.

© Springer-Verlag GmbH Germany, part of Springer Nature 2020
M. Jantzer et al., *The Art of Engineering Leadership*,
https://doi.org/10.1007/978-3-662-60384-0_1

When we created the leadership program we reviewed a large number of publications that deal with some of the aspects of engineering leadership. In our view today there is no holistic answer to the question: "How to develop a good engineering leader?" The extraordinary response and the overwhelmingly positive feedback of the participants encouraged us to write this book. It is a documentation of proven leadership practices. It is intended to encourage the conscious shaping of leadership roles. It supports leaders in quickly growing into a new role and shaping change in their organization.

The book is built-up of self-contained chapters that can be read without the context of the other chapters. The individual chapters are arranged in the order in which strategic goals are developed and implemented.

We start with the Why, the purpose of the organizational unit. We discuss the strategic targets of the units and the value contributions they provide. We explain how to secure the future viability of the organization by opening up new opportunities through innovations.

In the technical chapters like systems engineering, requirements engineering, architecture design and model-based development we explain essential elements for their suitable deployment by the leader.

In the subsequent chapters, we discuss how we implement quality attributes, realize development opportunities, systematically reduce risks and thus prevent creeping quality disasters. To do this, we need a suitable review culture, decision-making ability and strategies for dealing with complexity.

After clarifying what should be developed and which product characteristics and performance values should be achieved, we turn to the organization and the people in the last chapters. We discuss organizational design, roles, conflict resolution and building high-performance teams.

# Purpose of Leadership

**2**

▶ This chapter discusses our basic understanding of leadership in product engineering: leadership aims to shape and secure the future and strengthen the community.

We present a simple model for reflection on one's own leadership behavior, which accompanies us through all leadership situations.

## 2.1 Why is there Leadership?

Why do all societies form hierarchical leadership structures? What are the advantages for the members of a group? Since leadership structures are everywhere, they must serve some human purpose. We think that leadership supports two aspirations in particular:

- the shaping and safeguarding of the future of its members and
- the strengthening of their community

Thus, leadership has a service orientation to ensure the effectiveness and success of one's own group (see also [1]).

Let us first consider the shaping of our future. Developing a tangible picture of the future requires expertise, imagination, curiosity and confidence. If leaders share their feelings, they invite other people to follow.

Is there a link to engineering? Yes, because research and development are mainly working on a better future. Leaders in the company develop their picture, their ideas for the future in the strategy process. It starts at the highest operational goal of a company, the return on investment. In order to be able to operate on the market in the long term, the capital invested must generate a payback within a certain period. The decisive

© Springer-Verlag GmbH Germany, part of Springer Nature 2020
M. Jantzer et al., *The Art of Engineering Leadership*,
https://doi.org/10.1007/978-3-662-60384-0_2

question is: With what is this achieved? The "with what" is formulated by strategic objectives (Fig. 2.1). They illustrate the assumptions about the products and services that will provide a payback tomorrow on the capital invested today. Meaningful payback is linked to introducing something new to the market. Product development must therefore break new ground. How to do that? This is exactly where domain expertise comes to the fore. It enables leaders to recognize opportunities faster.

The development of strategic objectives is a central task of leadership. Strategic objectives provide purpose and orientation for product engineers. The strategic objectives anticipate the future. They make a future state tangible.

Development Engineers perceive operational targets, such as earnings and sales figures, rather as the result of successful work. They certainly enjoy high earnings and good sales figures, but meaningful strategic objectives are a much more powerful motivator.

To ensure that everyone is fully committed to the same strategy, it is important to coordinate and integrate all domains and roles involved. This brings us to the second goal of leadership, the strengthening of a common spirit and the formation of a strong community. There is an economic benefit to that. A strong community can achieve objectives more efficiently when everyone is moving in the same direction. This experience of shared and purposeful work gives people confidence, increases motivation, and thus indirectly promotes performance. Leadership does not have to provide motivation. People are intrinsically motivated to perform if they feel that they can contribute effectively. The role of leadership is to facilitate the creation of a community that engineers want to join. Community building leadership is most effective "from within the group".

Leadership helps the group to experience effectiveness by building trust and confidence. Leadership takes responsibility for the path into the future. This creation of purpose and orientation helps everybody in the group to gain courage and tackle the common engineering tasks. Leadership does not have to be hierarchical in any way. It can work on the same hierarchical level or even from bottom to top.

Now the engineers have a meaningful target and a path to the target that is almost visible. Is this already getting people moving? As social beings people do not want to go alone. They need companionship. They prefer to experience and master challenges together with a community. Applied to product development, this means that leadership

**Fig. 2.1** Strategic objectives and strategies

supports the engineers in deploying the strategy. Leaders remove impediments that appear on the way, close gaps on the way, and explores alternatives. Leadership shapes a culture of mutual and complementary support. In an agile context, the term "servant leadership" is used. As in the development of strategic objectives, the aim is to deploy the full potential of product engineers through trust and confidence, to increase the effectiveness of the group and of the individual. Therefore, especially in volatile situations, where nobody knows exactly what is right, agile methods prevail. Agile frameworks build on a culture of common and complementary work (see also [2]).

Leadership has two goals (Fig. 2.2):

- the setting of strong strategic objectives, or top-down leadership.
- strategy deployment, which means the support in implementation. This is also called servant leadership, it works bottom up.

However, leadership should not be understood hierarchically. Leadership is found at all levels and in all roles of product development.

Experts lead the further development of their domain. They enable others to do engineering according to the state-of-the-art. Top management develops the vision—an achievable long-term goal, the portfolio, the roadmap—and supports the lower levels in realizing it. Design engineers or software developers translate requirements into products and elaborate the feasible approach with colleagues in the team.

Leadership can be distributed over several roles. The agile framework roughly assigns the responsibility for development targets and results to the "product owner", the responsibility for realized solutions to the "team", and the formation of the team to the "agile master".

Each role, each hierarchy has its own task. Leadership in product development means connecting people with the solutions to be developed.

Leading engineering is about technical content and about the people who create it. We therefore speak of **leadership by content**, or **technical leadership**.

How does leadership differ from management? Leaders define the objectives, set boundaries and design the social system in which the engineers work. Leaders develop the work organization. Leaders in product engineering act as architects and chief engineers. Management acts inside the system, maintains it, keeps it alive and operates it.

**Fig. 2.2** Task and purpose of leadership in engineering

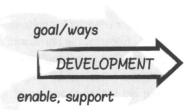

Management is an important activity. However, in this book we focus on leadership—the design of an engineering system—because management tasks seem to find managers automatically. Leadership requires reserved time and freedom to act. Leadership requires initiative and deliberate action.

## 2.2    Fundamental Aspects of Leadership Behavior

Everyone knows managers who have a very one-sided understanding of leadership. For example, the administrator, in whose organization everything is regulated perfectly, but nobody knows which goal the organization actually pursues. Or the visionary who creates wonderful images of the future but is not in a position to make a concrete decision.

A balanced understanding of leadership means mastering all tasks connected with leading people and technical topics. It also means recognizing the tasks correctly and reacting accordingly. Finally, when a team needs a decision from its leader, it does not want to be lectured on the decision making process or hear a flowery presentation on the future vision.

In our leadership trainings, we found leaders usually do not talk to each other about their own leadership behavior. In numerous discussions we have identified three main causes:

- Uncertainty about leadership behavior is seen as a weakness. Therefore, it is not discussed.
- Many colleagues regard leadership as something uniquely personal that cannot be objectively evaluated. A typical statement is: "I have my own style." This means that a discussion or feedback is not possible.
- Another reason is the lack of a common understanding about leadership tasks.

The mastery of leadership tasks is particularly important in product engineering. Leaders in product engineering have to deal with uncertainties, with competing goals, and with people operating within different motivational structures. In order to master these tasks, we need transparency, a common understanding of the tasks, and the willingness to learn together.

To facilitate a discussion about basic principles of leadership behavior in daily life, we condensed all leadership tasks into five behavioral dimensions, which are shown in Fig. 2.3.

"Act with integrity" emphasizes responsibility for building trust and building relationships. Integrity, the correspondence between declared values and perceivable action, builds trust. For most employees leadership action speaks louder than words. Injustices in treatment perceived by employees, dishonesty or a lack of transparency in communication are typical examples of alleged or actual deficiencies in integrity. In order to build up a sustainable, trusting relationship with employees, colleagues and leaders, it is

**Fig. 2.3** Basic principles of leadership behavior

essential that leaders walk the talk. Thus, for example, typical power games are a no-go, since they normally contradict the values of a company. The picture (Fig. 2.3) shows this leadership task as a person's leg, because it is a foundation on which the other tasks are based.

The second leg on which the leader stands is called "give sense of purpose". This means explaining the meaning and purpose of goals and tasks to others. When people do not understand the meaning and purpose, it is difficult to focus on the common goal. This leads to unproductive conflicts. At a low escalation level, these conflicts are expressed by a lack of understanding for the activities of others. For example: "I cannot understand why manufacturing promotes this outdated process." At a higher escalation stage, it might sound like: "Engineering has once again delivered a design requiring assembly forces that are too high". If we do not understand the underlying reason, we easily get the feeling that decisions could be based on competence deficits or even selfish motives. Therefore, the two legs, the clarification of meaning and purpose as well as acting with integrity are the basis for trust, community and motivation.

The management tasks "decide" and "organize" increase the effectiveness and efficiency of the organization, bringing delayed or incorrect decisions to zero. In this context, "decide" means above all the initiation of the decision and taking responsibility for it. This "arm" is so important that we have dedicated a chapter to it.

By "organize" we mean the shaping of processes and organization. Here, too, it is not about the exclusive design of an organization by the leader, but rather about the willingness to constantly look for new, suitable procedures and organizational structures and to implement them quickly. The quality of the implementation of these two leadership tasks has a decisive influence on the effectiveness and efficiency of an engineering organization.

The last leadership task, symbolized by the head, is to "inspire". We inspire the team for the common goals. This involves each individual employee, his/her individual motives and the ability of the leader to care for them individually.

Reflecting on leadership performance regularly in these five areas of behavior is a key to improving performance. If peers and superiors are included in it, leaders can supplement their view by the feedback of others. Leaders can first formulate their expectation on the leadership performance and then compare it again and again with colleagues, employees and superiors.

▶ **Practical Tips**
- Consciously plan time to reflect on your leadership performance.
- Use the ADIOS model to prepare for a meeting, for self-reflection at the end of a working day, or at the end of a meeting. What are the expectations towards you?
- What does integrity mean to you in your daily work? How do your employees and colleagues perceive you? Search for specific feedback based on the dimensions of the ADIOS model.

**The Most Important in Brief**

Leadership secures the future and strengthens the community.

The tasks of the leader are strategy development and support of their teams in strategy deployment.

Leadership by content connects people with development tasks in order to achieve targets.

With the ADIOS model we can reflect our leadership behavior in daily life simply and effectively.

## References

1. Sinek, S.: Leaders Eat Last. Why Some Teams Pull Together and Other Don't. Penguin, New York (2017)
2. Medinilla, A.: Agile Management. Leadership in an Agile Environment. Springer, Heidelberg (2012)

# Our Compass, or "Where is True North?"

**3**

▶ Using the Bosch product engineering system as an example, we explain the power of a "True North". It provides orientation in the dimensions of "results, procedure, leadership, skills and work organization" without specifying the path and procedure in detail.

We show that engineers not only comply with the law, but also contribute to society. This increases the sense of purpose and can lead to high performance.

With direct leadership, managers can always ensure that important issues—whether product features or development approaches—are implemented in the way they want. However, in this case leadership may become a bottleneck for engineering. The leader's capabilities determine the pace and quality of development.

If you want to achieve more, you have to give the responsibility to the teams. However, how do you now ensure that your priorities are maintained?

One possibility is to formulate a "True North"—a target state that you want to achieve. North is always where the compass needle points. The "True North" aligns the compass for the engineers. It remains stable over relevant time scales. When your employees are heading north, it is not so important that they follow exactly the same path. It is important that they move towards north—even if they do not take the direct route.

We have created a product engineering system for Bosch that serves as a reference for good engineering. Our expectation are not only excellent results, but also an excellent approach and execution.

© Springer-Verlag GmbH Germany, part of Springer Nature 2020
M. Jantzer et al., *The Art of Engineering Leadership*,
https://doi.org/10.1007/978-3-662-60384-0_3

## 3.1    The Bosch Product Engineering System

### Vision and Goal

---

**Passion for Engineering**

We employ our creativity and courage to break new ground. We develop fascinating products which deliver outstanding quality at attractive costs.

---

There is always a tension between innovative products with high customer benefit and cost. Successfully navigating this conflict requires engineers with passion. It also requires leaders with courage and confidence. The market will not accept attractive new solutions that are simply too expensive. Take for example Apple's Lisa in 1983, which cost US\$ 10,000 and was just too expensive for a PC, even with the new and very attractive mouse. However, the mouse became a standard—at attractive costs later.

### Principles

The Bosch product engineering system is based on the following five principles. There are just five so that they are easy to remember.

1. create value
2. understand products
3. lead by content
4. strengthen competences
5. work smart and agile

The principles illustrate our mutual expectation in a highly condensed form. It's what we are measured by.

To make it more precise and applicable we added a short explanation to each principle. For example on "work smart and agile":

---

**Example**

*Work smart and agile: Evaluating the type of tasks, we actively shape the cross-functional and international collaboration using agile and lean principles. We ensure clarity of tasks and transparency in workflows.*

---

This is our true north, our yardstick.

### The Engineering System

The above principles span different dimensions: Results, approach, leadership, skills and work organization.

How are they connected, how should it be understood? This is outlined in the development system (Fig. 3.1, Elements).

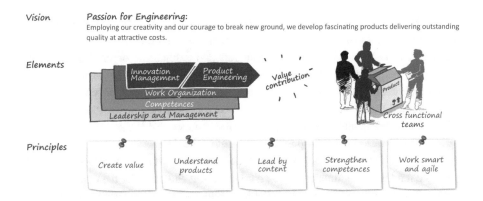

Vision   **Passion for Engineering:**
Employing our creativity and our courage to break new ground, we develop fascinating products delivering outstanding quality at attractive costs.

**Fig. 3.1**  Bosch Product Engineering System

The principle of "create value" is the goal of every development. Engineering wants to create valuable products and services for customers and users. Several functions are usually involved in this. In addition to engineering, we generally need purchasing, sales, production and other functions. We always see work results from engineering as a contribution to the value provided to the customer by all functions: a value contribution. This strengthens the awareness that we must work together—for the benefit of the customers. It is intended to reduce silo mentality of functional units.

The generation of a value contribution by engineering can be described by its value stream, which begins with innovation and is driven to market maturity through product engineering. We are constantly striving to optimize that value stream and integrate even more effective procedures. Because it is this value stream that ultimately leads to success. That is why we have placed it at the top of the product engineering system.

The work organization supports the value stream. It does not create value itself. Therefore, the work organization follows the task (see the principle: work smart and agile). The work organization must remain as flexible as the value contribution may change over time. Although work organization is based on tasks, this element starts somewhat earlier than the value stream, because work has to be (roughly) organized before it can be carried out.

In addition to the value stream, the work organization is also based on the skills available in the organization. Of course, skills need to be developed just like the value contribution we want to create. Therefore, the element of developing skills starts earlier than the work organization. Although we also acquire skills during value creation, we want to develop the competences beforehand. This is more effective and efficient.

The following may be unusual for many leaders: Leadership and management are placed at the very bottom! Leadership is positioned at the beginning of all work in our system. We start the development with the formulation of strategic objectives. Then, leaders develop the other fields of the system and shape the elements. They tailor and deploy the product engineering system in their area of responsibility. That is why the field of leadership and management starts earlier than the other elements.

Once the leaders have developed the strategies and defined the organizational structure and processes, it is their task to bring the entire engineering system to life. They create optimal conditions for value creation. Leadership ensures success, removes impediments, and overcomes resistance. This is a service function for those who provide the operational results and create value: the engineering teams and the experts. Regarding value creation leadership is servant most of the time.

As it is very important to us, we have added another picture: The cross-functional team. People tend to overestimate their own contribution and consider the contribution of others to be less important. To recognize the importance of our development partners, we gave cross-functional cooperation a key visual. We are convinced that the optimum product is created in a smooth collaboration across all functions.

**The Visual of "True North"**
All in all, our true north looks like this (Fig. 3.1).

It fits on one slide and outlines what we are striving for. It leads to good discussions within leadership trainings. Our teams can act empowered, guided by a clear understanding of the "true north".

## 3.2   Social Responsibility

Today, it is not enough for engineers to limit their own responsibility to compliance with the law. Behind laws are expectations and needs of users and the society. They are usually not explicitly mentioned in the legal text.

Thus the VDI ("Verein Deutscher Ingenieure" or "Association of German Engineers") writes in its preamble to the "Fundamentals of Engineering Ethics" [1]:

**Example**
Engineers recognize natural sciences and engineering as important powers shaping society and human life today and tomorrow. Therefore, engineers are aware of their specific responsibility. They orient their professional actions towards fundamentals and criteria of ethics and implement them into practice. The fundamentals suggested here offer such orientation and support for engineers as they are confronted with conflicting professional responsibilities.

Leaders also have the task to align decisions with the underlying expectations of users and society. This may create conflicts, as meeting additional expectations can lead to increased product costs. This conflict reveals the purpose of engineering, i.e. to balance economic and social interests. Recognizing this purpose can become an effective enabler in the hands of leadership. Making a contribution to society provides deeper meaning to our work as engineers.

As engineers, we play a key role in shaping people's environment. An engineer who develops a new washing machine makes life easier for users while protecting natural resources. Engineers in vehicle technology, for example, are working to reduce road traffic fatalities. They develop sensors and actuators that enable safe braking, stability control, or airbags that effectively protect vehicle occupants in the event of an accident.

As social beings we want to make a contribution to the community. This inspires us and drives us to top performance. Increasing the growth and earnings of your own company is not enough to provide purpose. There is more to it. Bosch has summarized this requirement in "Invented for life".

▶ **Practical Tips**
- Exchange information with your colleagues about success factors in your engineering system. Derive your principles from this.
- Discuss with your colleagues the overall contribution of your work to society.

---

**The Most Important in Brief**

The "True North" gives orientation without defining every step of the way.

Shared principles create a common identity.

Engineers are not only committed to their company's earnings targets, they also contribute to society.

---

# Reference

1. Ethische Grundsätze des Ingenieurberufs. VDI, Düsseldorf (2002)

# Strategic Objectives

# 4

> We explain the importance of strategic objectives using an example. We show their derivation from a consciously chosen future scenario and discuss success factors and typical difficulties. We present tools that facilitate the development of a strategy, i.e. concrete steps towards achieving the objectives.

"All men can see the tactics with which I conquer, but no one can see the strategy from which victory arises."—From: Sunzi: "The Art of War" around 500 BC [1].

In our work with engineering leaders, we often receive cloudy statements about customer orientation, innovation and efficiency, about our claim to become market leader, and sometimes a reference to the corporate strategy when asked: "What is the strategic objective and strategy in your area of responsibility?"

Since these answers are unsuitable as strategic objectives and rather the sign for a missing verbalized strategy, we will clarify the question "Why do companies and development teams need strategic objectives?" For this, we put ourselves in the position of Klaus Maier.

---

**Example**

Klaus Maier is the head of engineering at a manufacturer of self-tapping screws. The company' objective is to increase sales and profits by 30% over the next 5 years. His task is to advise medium-sized customers on the design and selection of suitable screws for special load situations. With the experience and knowledge of his team, Mr. Maier has earned the trust of his customers.

In this short description, we find a picture that is oriented towards today. The company produces screws for special applications for medium-sized customers. Through consulting and design, the company delivers a value contribution. It has thus gained a sustainable competitive advantage. The current strategy does not take into account

---

© Springer-Verlag GmbH Germany, part of Springer Nature 2020
M. Jantzer et al., *The Art of Engineering Leadership*,
https://doi.org/10.1007/978-3-662-60384-0_4

any market changes and therefore does not explain how the current market leadership is to be strengthened in the coming years.

Let us continue to imagine the following scenario:

In recent years, Mr. Maier's customers have started using snap-fits made of injection-molded plastic. After initial difficulties, the technology has now matured to such an extent that snap-fits can replace self-tapping screws in relevant areas.

This poses a threat to Mr. Maier's business. There is a threat of a partial replacement of his current product technology. Certainly, there are also new market opportunities.

Mr. Maier decides to invest part of his team's resources to focus on the future:

He thinks: To continue as before could lead to a decline in sales. A simplistic approach like "We will also develop snap-fits" would fail because the company has no strengths in this area. There is no business model available for these kind of activities. He asks: "Which of our strengths can help us to win in the future? What problems do our customers have that we could solve for them? What are the hurdles to using our products in other areas?"

Because of these challenges and the large number of opportunities, engineering organizations need strategic objectives for their future business, and plans to achieve them.

## 4.1    How Do We Develop Strategic Objectives and Strategies?

In our example, we have seen that a strategy is first a plan to meet a challenge. This is associated with uncertainty. The plan is based on a general hypothesis: "Because I expect this threat or opportunity to materialize, I believe that this is how we will do business in the future".

Let us look again at Mr. Maier:

**Example**

Mr. Maier and his team analyze the situation of their customers and their own competitive situation.

Mr. Maier focuses on the strategic objectives defined by his company. They provide search directions or guidelines. The company wants to grow with internet-based services and offers. Mr. Maier and his team develop scenarios with different business models. They decide to attract new customers by product and process engineering with a web-based design and ordering app. In his imagination, Mr. Maier already sees how his customers parameterize their screws in the web app with just a few clicks and then place the order. Mr. Maier formulates his strategic objective as follows:

"We enable our customers to accelerate their own development process via an intuitive platform for designing and ordering self-tapping screws. This means that we will achieve growth of 37% in machine building and plant construction sector in Germany and Europe over the next three years and increase our EBIT to 9%."

We see that Mr. Maier has found his strategic objective. Based on a solid analysis of his environment, and the strengths and weaknesses of his team, he determined a desirable future state, which leverages the strengths of his team. He has changed his perspective to consider not only his own value creation, but also the entire value stream from the perspective of different customers. He analyzed the strategy of his competitors and formulated a new value proposition.

The objective describes a future business model. It is concrete and inspiring. It contains a value hypothesis and it is realistically achievable. Based on this objective a concrete planning is possible. Mr. Maier and his employees can now develop different strategies to realize this business model.

Developing a strategic objective requires openness and transparency. This is intentional because it stimulates discussion and competition for the best idea. The task of leadership is to facilitate this discussion in order to achieve the best result for the community.

This simple example shows: Task descriptions are by no means enough, as they often focus solely on the present. Global objectives, whether operational (30% sales and profit growth p.a.) or strategic ("We will become the most customer-friendly provider") are inspiring at best. Engineering leaders need to be more specific and guide the creativity of the rare and expensive resource of engineers.

## 4.2    Success Factors for Deriving Strategic Objectives

In our discussions with engineering leaders, we occasionally encounter poorly formulated strategic objectives. Why is that so?—Let us look at the typical mistakes made in the development of a strategy and the definition of a strategic objective.

We have already indicated the most frequent mistake: focusing only on short and medium-term tasks. Instead, it is essential to use upper level strategic objectives as a guardrail and to search for a future value proposition. A good strategic objective stimulates discussion. If everyone can only say "yes", it is too superficial.

Strategic objectives that are so far away that they seem unattainable do not provide the necessary orientation. Realistic intermediate objectives may help. For example, car manufacturers have divided the objective of autonomous driving into several levels, ranging from automatic distance adjustment to fully automatic driving (levels 3–5). This allows the engineers to orient themselves and derive strategic objectives that describe their value proposition as a part of the puzzle.

The most important success factor is the willingness of leaders to make themselves vulnerable by translating higher-level organizational objectives into specific strategic objectives and measures for their team. This demonstrates commitment to strategic objectives of the company. Without leadership's identification and commitment to a strategy, employees do not pursue the objectives set for them.

Fig. 4.1 PESTEL-analysis
tool to identify challenges

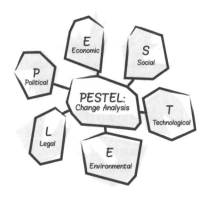

## 4.3    Tools for Strategy Description

In our small example, we have drawn the picture of an entrepreneur making a "lonely" decision. In a company's environment, such a "lonely" approach may or may not work. To be successful, employees, colleagues and superiors must be involved. A common view of one's situation, opportunities and threats is essential to inspire everyone to action. The following tools are available for this purpose:

The PESTEL analysis (Fig. 4.1) [2] is a suitable tool to analyze threats and opportunities. PESTEL stands for Political, Economic, Sociological, Technological, Environmental and Legal change. This is a kind of checklist to identify changes in the relevant context. The list is designed to provide a consistent perspective on all the challenges (threats and opportunities) within an organization. Based on this collection, leaders can prioritize and focus.

A compact view of the strengths and weaknesses of your company and your competitors is obtained by means of a SWOT analysis (Fig. 4.2). SWOT stands for Strengths, Weaknesses, Opportunities and Threats. This view serves to assess the organization's ability to pursue a specific strategy.

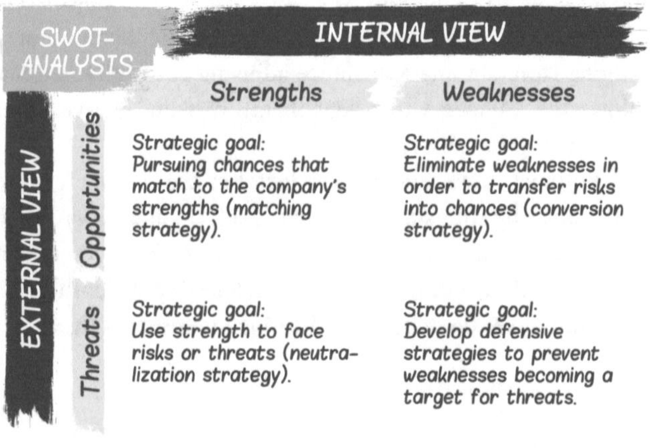

Fig. 4.2 SWOT analysis
links chances and risks to my
specific situation

| SWOT-ANALYSIS | | INTERNAL VIEW | |
|---|---|---|---|
| | | Strengths | Weaknesses |
| EXTERNAL VIEW | Opportunities | Strategic goal: Pursuing chances that match to the company's strengths (matching strategy). | Strategic goal: Eliminate weaknesses in order to transfer risks into chances (conversion strategy). |
| | Threats | Strategic goal: Use strength to face risks or threats (neutra-lization strategy). | Strategic goal: Develop defensive strategies to prevent weaknesses becoming a target for threats. |

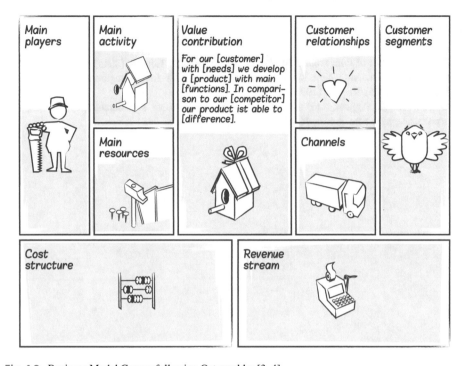

**Fig. 4.3**   Business Model Canvas following Osterwalder [3, 4]

A common way to make a business model transparent is the "Business Model Canvas" (Fig. 4.3) [3]. With the value proposition at its core, it reduces the complexity of a real business model to the essential dimensions. Leaders and experts can discuss and compare different business models at the same level of detail.

▶ **Practical Tips**
  - Study the strategy of your company and derive the guidelines for your own strategy development. Determine your vision of the future for your area of responsibility. Utilize your freedom to derive suitable targets and deduct contributions for your own area of responsibility
  - Discuss your results with your colleagues, your boss and your employees in order to gain commitment.

---

**The Most Important in Brief**

A strategy is based on a possible future scenario (hypothesis) and derives concrete objectives from it. The strategy as a path to reach the objectives firstly describes the identity and capabilities of your own company and secondly identifies the necessary additional skills. Concrete steps to achieve these objectives are defined.

# References

1. Sun Tzu: The Art of War, around 500 B.C. Penguin Classics, London (2015)
2. Ginter, P.M., Duncan, W.J.: Macroenvironmental analysis for strategic management. Long Range Plan. **23**(6), 91–100 (1990)
3. Osterwalder, A., Pigneur, Y.: Business Model Generation. Wiley, New Jersey (2010)
4. www.strategyzer.com. Accessed 7 Nov. 2018

# Creating Value—Providing Sense of Purpose

**5**

▶ In this chapter we introduce the concept of value contribution. Value contributions are created for the customers and users of products.

We show how leaders and engineering teams can achieve clarity about their own value contribution.

Why do customers buy your products? What value do your customers see in your products today? Moreover, will your product still have value for them in the future?

Companies often focus on optimizing and refining product functions. Their innovations aim to implement the same functions even faster, better or more cost-effective. But the underlying value of the product for specific customers may differ from those functions. If technical alternatives get ready for the market, the focus on these functions may even be detrimental for future business. Sustainable success depends on understanding and optimizing the customer value, which means providing the customer new abilities or new opportunities. This includes emotional aspects such as pride, prestige or spontaneous enthusiasm. Let's look at an example:

---

**Example**

Kodak has produced cameras, films and photographic paper for many years and was the world market leader for photochemical films. Users could easily take pictures. However, they could only view the photos after a complicated and time-consuming multi-stage chemical process. Nevertheless customers bought Kodak films because of their excellent color reproduction and good contrast.

In 1975, Kodak engineer Steven J. Sasson invented the digital camera. A first model was launched in 1991, but it was not suitable for the mass market. Kodak continued to optimize analog photography as a core technology. Other companies

---

© Springer-Verlag GmbH Germany, part of Springer Nature 2020
M. Jantzer et al., *The Art of Engineering Leadership*,
https://doi.org/10.1007/978-3-662-60384-0_5

meanwhile worked on digital photography, bringing models with increasingly better resolution to the market.

With digital cameras, customers could preserve their memories just as perfectly as with the old wet-chemical images. They could also see the pictures immediately and show them to others. The first cameras reproduced the colors unrealistically: skin tones appeared waxy and grass was frog-green. But the hobby and professional photographers preferred to evaluate their pictures immediately, edit them on the computer, and show them to others. Even the much worse resolution at the beginning was not an obstacle. A new value proposition that was highly attractive for the photographers had been created.

Although the technology was available within the company, Kodak did not drive the change. It was not until 2004 that Kodak started to focus on digital photography. By this time, however, other companies had already gained a leading position. Between 2004 and 2012, Kodak gradually discontinued camera and film production and sold its core business in 2013.

What is the take-away from this example? New technologies provide new abilities to the customer. This may fundamentally change the value that consumers attribute to existing products or services. Leaders have to face this challenge repeatedly and align themselves and the business accordingly.

## 5.1    Value Contribution

The product alone rarely provides the customer value. In the example of the camera, good instructions, service offers, suitable accessories, and attractive packaging provide additional value to the customer.

The customer journey is a useful tool to identify these values in a systematic way, see Fig. 5.1. It helps to identify all touch points with the customer that contribute to the value. The requirements at each touch point are identified and incorporated into the requirements for the overall product (see Chap. 8).

Today, value is created in organizations with a high degree of labor specialization. Engineering is divided into many disciplines (software, hardware and even specialty competence centers). Each discipline strives for perfection and thus, individual functions and features are constantly improved. It is rather easy to loose sight of the real customer value.

Engineering leaders have the task to synchronize their organizational units with the others and to align them for best customer value. Technological change may require other collaboration models. Precision engineers suddenly have to align with software developers to network a camera. If this even results in organizational changes, leaders should provide a convincing change story.

**Fig. 5.1** The product journey shows all touch points that contribute to the value proposition of the product

product information  purchase  installation & configuration  product use  maintainance & support

In our leadership seminars, we use the expression of "contributions" to customer value (of the overall offer). We start the description of the value contribution from the point of view of the customer. What ability does the customer achieve through our contribution? How great is his appreciation (Fig. 5.2)? Each organizational unit delivers partial contributions with its competences. As leaders exchange their views, they are able to align. For internal cooperation, it is beneficial to discuss mutual expectations and feedback on the delivered work result. At the same time, they orient their employees. We regularly experience that many employees are unaware of their contribution to the big picture. But as soon as it becomes tangible, it has a very motivating effect.

**Fig. 5.2** The value contribution

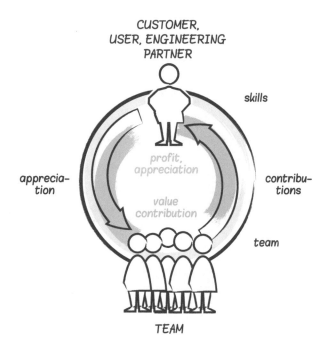

CUSTOMER, USER, ENGINEERING PARTNER

skills

appreciation

profit, appreciation

value contribution

contributions

team

TEAM

The leadership instrument of value contribution thus serves the holistic orientation towards customer value. If an organization orients itself on customer value, it can adapt quickly to changing customer value and technologies.

▶   **Practical Tips**
- Describe the value contribution for yourself and your team. Take the customer's point of view consistently. Focus on your core contribution and avoid describing your tasks.
- Make it clear to yourself and your team what inspires your customers and how their appreciation is expressed.

**The Most Important in Brief**

Product engineering aims to create value for customers and users. Describing value contributions promotes development geared to customer needs. Value contributions are product features or services for customers. Appreciation can also be expressed in emotions such as enthusiasm.

Leadership must ensure that functions and services always align with user and customer value. The focus on user and customer value has a meaningful and motivating effect. It makes an organization efficient and adaptable to external changes.

# Systems Engineering

**6**

> This chapter describes the basics of systems engineering.
>
> We present methods such as decomposition, continuous integration and visualization using different views to guide system design.
>
> Trust is a powerful currency and we demonstrate its importance to systems engineering.

Which design tasks are typically complicated? If the scope of the task is big and vague and many disciplines are required to solve it, things get complicated. If this is not the case, design tasks can be tackled and solved directly.

Let us look at the development of complex technical systems in which many engineering disciplines are involved. The main function of an ESP® system is to realize the steering and deceleration commands of the driver in the best possible way in extreme driving situations. To do this, the system can control the deceleration of individual wheels or reduce engine power. It consists of mechanics, hydraulics, electronics, sensors and various control algorithms, which are implemented in software.

How to decide which functions are allocated to each engineering domain? How are all functions coordinated? Who defines the interfaces? Who distributes the design tasks? Is one discipline in charge? Are decisions made by consensus across all disciplines?

Because it is not so easy to decide what is right or advantageous over all requirements of a system, a new engineering discipline has emerged since the beginning of the 1990s: Systems Engineering.

© Springer-Verlag GmbH Germany, part of Springer Nature 2020
M. Jantzer et al., *The Art of Engineering Leadership*,
https://doi.org/10.1007/978-3-662-60384-0_6

## 6.1    Structuring Systems

Systems Engineering is the discipline of systems design [1, 2]. Systems Engineering is not "committed" to any single discipline, but works with all disciplines to develop the best solution to serve customer and stakeholder needs. Systems Engineering starts with the first idea to develop something and always considers the entire product life cycle. In addition to product usage (function), Systems Engineering always considers:

- costs (development costs, product costs, total cost of ownership, life cycle cost)
- schedule
- training and support of users and service
- desired product properties (quality)
- manufacturing, including purchasing and sales
- phase-out at the end of the product's life.

That sounds a bit unmanageable at first. Without a methodical approach it would be true. However, Systems Engineering models the system in all necessary views, not only functionally, but also in all dimensions in which decisions have to be made. This enables two basic principles to find comprehensive solutions:

- decomposition—also known as "separation of concerns" and
- continuous integration.

Decomposition means the break down of a system into manageable units. One differentiates roughly:

- Partitioning, the division of a system into smaller (sub)systems
- Domain-specific decomposition, e.g. noise vibration harshness or electromagnetic characteristics or costs
- The decomposition along the functional structure, characteristic structure or solution structure
- The decomposition along the timeline.

Depending on the problem at hand, a suitable decomposition to a manageable scope should be selected. At this level, alternative solutions are developed. This has the advantage that the problem to be solved is smaller, at least for the moment.

Once you have found a solution at that level of detail, you must of course check whether you can still achieve all targets of the project. To do this, you must reintegrate it into the overall problem. Integrate it into your overall system model as soon as the solution is ready for discussion and as often as possible—ideally continuously.

When you achieve all project targets and all major domains agree to the solution, the problem is solved. If the integration process reveals that the proposed solution has unacceptable effects on other project objectives, the next iteration is required. In this way, solutions are developed step by step or—according to agile approaches—iteratively and incrementally.

Systems Engineering is a structured and methodical development process (Fig. 6.1) that is feasible in any kind of work organization. It follows a logical sequence.

Every product development should start with the formulation of a strategic goal and strategic planning. They lead to the individual development projects that are directly or indirectly shaped by stakeholders.

The engineering team can start as soon as the project's core objectives and guidelines have been defined.

Requirements development, architectural design, and component development follow. These three steps that logically build on each other, cannot be meaningfully separated. Since later steps have an implicit retroactive effect on the preceding steps these design steps and their increments are also created iteratively.

This is followed by stepwise integration. Here the team validates and verifies whether the elaborated increments solve the problem. By doing this, they check the individual increments for compliance with the "definition of done"[1] and the "acceptance criteria"[2] of the project. These criteria are defined during requirements development. The results are logical relationships between the descending and ascending elements of the V-model. And, for sure, the team will learn something new during system integration, which will feed back into the requirements. Thus there are also large iterations over the whole V (Fig. 6.2).

When the system is ready and goes into operation, hopefully everything has been done right. Experience shows that users of new products sometimes want to do more with the product than the product engineers had in mind. Therefore, the engineers learn something new during roll-out and have to be prepared to react swiftly and appropriately. If usage data is recorded this reaction can be faster and more targeted.

Product engineering means learning—step by step. During a development the team has to accept that some unexpected learnings will happen until the very end. This incremental learning fits naturally to the incrementally approach of agile development. In an agile approach, the V-model is completed in full or in meaningful parts in each sprint. Major project stages might integrate several sprints.

In maturity-oriented product engineering processes, several parallel and (loosely) coordinated V-models are running in parallel per sample stage (concurrent engineering). At the end all parallel lines of action are integrated.

---

[1]General criteria that must be fulfilled to consider a development task as done.

[2]Task specific criteria that must be fulfilled to consider a development task as done.

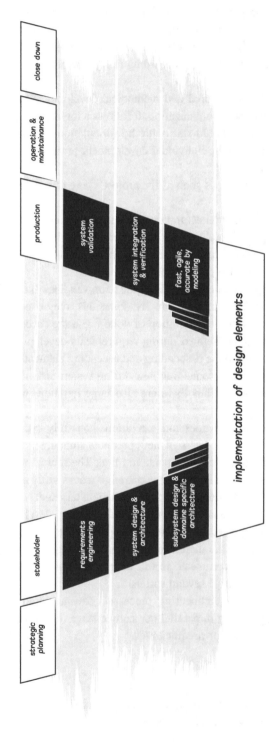

Fig. 6.1  V-model of systems engineering, following [1, 3, 4]

**Fig. 6.2** Working with the V-model

Decomposition & Integration          Iterative & Incremental          Verification & Validation

In both workflows, development is complete when all project targets have been achieved.

## 6.2   Currency in the Systems Engineering Team: Trust

Product engineering always operates in the area of uncertainty. Conflicts of interest must be resolved. How do you decide whether the self-test of an ESP® system is realized in software or integrated in the control chip? What does software cost over its life cycle and how much would the mask for chip production cost in addition? Do you need to stay flexible or not? These questions can lead to tough discussions. Can your company decide on the basis of life cycle costs or does it decide on the basis of piece part costs? In the second case, the result is clear: The self-test will be realized in software.

Changing the decision criteria would potentially change the outcome. Decision criteria are a big lever for leadership and therefore any change usually faces resistance. If the entrepreneurial target changes, e.g. from a component supplier to a software company, it is very likely that decision criteria will have to be adapted.

Systems Engineering will unfold its full potential in a corporate culture of open exchange, mutual trust and early decision-making by teams and experts who have access to all required information (Fig. 6.3). It strengthens the technical leadership in product engineering. Since the different domains can only know and work on parts of the "truth," a new role is required to integrate these many "truths" into one holistic picture.

**Fig. 6.3** Source of data and levels for decision-making

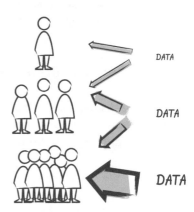

DATA

DATA

DATA

The task of the Systems Engineer is to develop the "best" comprehensive solution strategy. This gives him/her a special role with a lot of technical and business responsibility. Because he/she works with many engineers from many domains, it is very helpful for the Systems Engineer to work with domain-specific decision-makers in a way that strengthens mutual trust and integrates the respective strengths of the domains. This leads to complementary cooperation and responsibility between line managers, project managers and experts.

A Systems Engineer has formed a good team when the individual members can offer solutions for common problems from their domain and credibly present where their limits are and what disadvantages may arise. If solutions are expected mainly from other domains, the team has not achieved its optimum (compare Chap. 19).

For successful cooperation across domains, hierarchies and even across national borders—one thing is crucial: mutual trust and openness.

Why is it economically sensible to establish trust?

For two reasons:

- The target is to win on the market and not against other domains within the company. The less energy you have to invest in internal company disputes, the more energy can be used for the customer. You can concentrate on value creation and customer benefit. Everyone wins. It goes without saying that "we" can only succeed with real trust in colleagues, employees and managers. Mutual trust makes all involved parties more effective.
- Leadership should not slow down development, but inspire it. Collecting and evaluating all data, controlling everything is a lot of work. Especially when the task to be solved is complex or complicated. A well-coordinated team of experts and managers can do this most effectively. If managers don't have to question and test everything, they gain speed. Speed provides a competitive advantage. Trust therefore makes development faster, more agile and more cost effective.

But trust does not just appear, it must be built actively. Trusting someone without a profound basis is unprofessional. Both the manager and the expert would not live up to their responsibilities. Mutual trust is earned by performance (compare Chap. 19).

## 6.3   Enhance Ability for Sustainable Decisions

One of the decisive processes during the development of a product or service is the elicitation of requirements. This is well known by all engineering leaders and yet it is one of the most difficult technical tasks. Have you ever experienced receiving hundreds of pages of requirements on your desk. On it you find a little sticky note saying "Please release!". What to do now?

A few hundred pages cannot easily be checked for correctness and consistency. Is it not the task of the team to check that? The task of leaders is to commission "the right thing". To be able to decide these leaders need to understand the value contribution for the customer and the function of the next higher system level. It makes a difference to decide to raise the general temperature requirement to 140 °C or if a special function leads to self-heating to 140 °C. Leaders can only be effective in their role if the data is prepared in such a way that they can decide on the benefits for customers and the effects on stakeholder needs. Then leaders can contribute to problem solving, making portfolio decisions, discussing effort and prices with customers, and revising the strategy with colleagues and superiors. That is how leaders contribute to complexity management and thus support the team.

Structuring requirements hierarchically is key to effective decision-making. At the same time you can strengthen mutual trust in colleagues and experts. It also enables decision making on every level of the system. While the engineer needs detailed information for the design of components, leaders need use cases and user stories to be able to decide. In the end, both must be consistent and congruently modeled. The information on which decisions are to be made is prepared on each hierarchy level of the system (Fig. 6.4). Instead of hundreds of unstructured pages, horizontally and vertically structured documents with a manageable number of requirements per element are created.

Systems Engineering also follows the principle of "separation of concerns" and at the same time continuous integration. Completeness of the specification is achieved by systematically collecting, translating and partitioning stakeholder requirements.

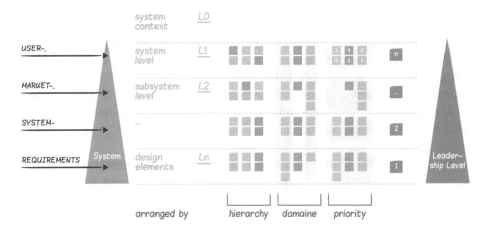

**Fig. 6.4** Hierarchical structuring of requirements

- Quality and legal requirements are derived from laws and industry-specific standards (in Fig. 6.4 blue squares).
- Requirements of users and the next higher or neighboring system are derived from analyses of user needs (in Fig. 6.4 green squares) as identified in the Customer Journey (Sect. 5.1) and from the technical analysis of the interaction with further systems.
- Business requirements are elaborated together with all relevant stakeholders (in Fig. 6.4 orange squares).

These holistic and hierarchically structured requirements also create an understanding of the essential value contributions for users and customers. The associated transparency promotes the ability to make decisions, to focus on what is important, and finally to support mutual trust.

How do you really do that? It may feel like too many interwoven expectations. Complex questions need to be structured first. One of the engineer's standard solutions is modeling, in this case modeling the requirements. Today this is typically done using SysML based tools. They enable networking and hierarchically structuring requirements. They also support the partitioning of requirements deriving problem-specific views (Fig. 6.5) from the overall model ("separation of concerns").

Specific questions can thus be discussed and solved in a focused manner. At the same time, the result can always be reflected holistically.

Which views are used to design the reliability of a sensor in a transmission unit (Table 6.1)? Different views from other disciplines as well as views from design for reliability are used.

The table shows how the question of reliability is solved step by step (in italics), taking into account function, cost and development effort.

The example also shows that the processing of such questions is only possible cross-functionally. Systems Engineering promotes rapid mutual understanding through mutual transparency, focusing on customer value and mutual trust, because the "why" for a given requirement can be clearly derived at any time.

Our leadership training courses focus less on the technical complexity associated with Systems Engineering than on interpersonal skills at all levels. Intercultural and interdisciplinary communication, the "right" work organization, the "right" decision, and the "right" approach to come to decisions are emphasized.

Even if you are technically on track, you can fail due to interpersonal complexity. At the risk of being repetitive, the key to success is trust. Why? You listen better to people you trust. You understand their concerns better. You also can trust not to be exploited. You do not have to put energy into self-protection. In the event of a conflict, a much larger solution space is opened. Rather than being left with a (rotten) compromise, an environment of trust generates new solutions.

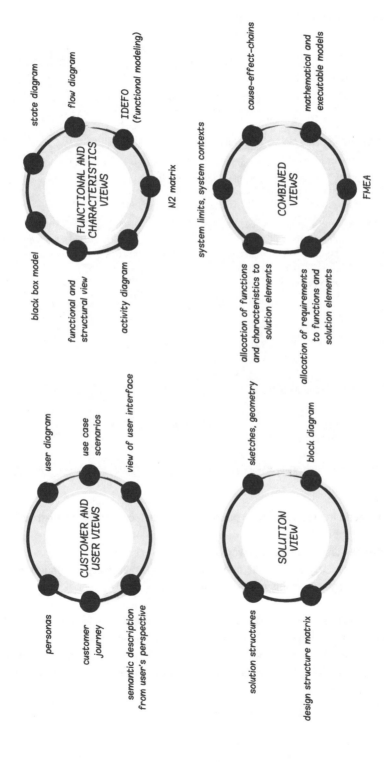

Fig. 6.5   Selected views of systems engineering

**Table 6.1** Exemplary views of reliability design

| Tasks | View |
| --- | --- |
| The required function of the sensor is derived from the system function of the transmission control. In this case, this is the detection of the rotational speed of a shaft | System design |
| The speed range and the required accuracy are derived from control engineering | Control logic |
| The operating conditions (media, vibration, temperature, operating time…) result from the next higher system (the internal loads of the gear unit) and a user analysis (external loads on the gear unit). In addition, the loads during production and logistics are also considered | *Load paths,* user stories, geometry, production process, logistics |
| Next, the responsible management decides how many gear units are allowed to fail over their nominal service life. This requires a model for the users' quality expectations. It must be considered what the consequences of a failure are and whether there may be legal requirements (over the intended product life cycle). The accepted number of failures depends on the product area and is different for an electric whisk than for a passenger car. This figure must be determined with care, as it determines both cost and quality perception in the market | User Experience, standards, laws, quality, *system reliability,* costs, development time and effort |
| When the permissible failure probability has been determined at the transmission level, the reliability target can be partitioned from the system level to the components in the system | *System partitioning,* costs, statistics |
| Verification of the partitioning at system level: Have all required value contributions to the customer been realized? Are all stakeholder requirements taken into account? Is partitioning robust against fluctuations of specification? What additional opportunities and threats exist? | Verification, opportunities, risks |

What can you do in everyday life to build up trust? Reduce sources of mistrust.

- Do not assume things are clear anyway.
- Use terms as precisely as possible and do not assume that everyone understands the same thing. Different domains use the same words differently.

Misunderstandings promote mistrust. Understanding promotes the joint search for solutions, unleashes the creative potential to solve problems in a better way.

Engineering science has developed a strong means of communication since very early times: The model—a simplified abstraction of reality (compare Chap. 10). It is also inherent to Systems Engineering.

▶ **Practical Tips**

- Allocate the system elements of your area of responsibility within the higher-level system and define the input/output relationships at the system interfaces.
- Determine your scope for design and decision-making and formulate your expectations to others.
- Agree on the views to be used in your work area and demand their consistent use.
- Check whether your team has sufficient competence in systems engineering.
- Strengthen mutual trust in your teams and actively build trust.

---

**The Most Important in Brief**

Systems Engineering is a modular methodology for the development of complex systems.

Well-structured and partitioned requirements and the use of agreed views improve cooperation and enable efficient decision-making.

The best possible solutions can only be found in a culture of mutual trust.

---

# References

1. David, D., Walden, G.J., Rödler, K.J., Forsberg, R.: Douglas Hamelin, Thomas M. Shortell: Systems Engineering Handbook. Wiley, New Jersey (2015)
2. NASA Systems Engineering Handbook. NASA/SP-2016-6105 REV 2. NASA, Washington D.C. (2016)
3. VDI 2206: Entwicklungsmethodik für mechatronische Systeme – Design methodology for mechatronic systems. Beuth, Berlin (2004)
4. Systems Engineering Guidebook for Intelligent Transportation Systems, Version 3. U.S. Department of Transportation. www.fhwa.dot.gov/cadiv/segb/files/segbversion3.pdf (2009). Accessed 21 Sept. 2018

# Leading to Innovation

**7**

▶ This chapter explains the difference between an idea and an innovation and how innovation is promoted. Leaders create boundary conditions and prepare the ground for innovations.

We present approaches to identify user needs and find new approaches to solutions.

Decisions have to be made continuously, even in uncertain situations. Often sceptics have to be convinced. Courage to break new ground is necessary.

## 7.1    When is a Good Idea an Innovation?

Innovations are new solutions that have been successfully introduced to the market. A lot more than a good idea is needed to create an innovation. To illustrate this, we will show how not to do it. Some managers believe that it is sufficient to organize creative workshops. For this purpose, developers must leave the project's daily work and join others for a day or two. A facilitator runs different creativity methods. At the end of the workshop, ideas are quickly evaluated and presented to management. Perhaps individual issues will be followed up further. Then real life returns. In daily routine new ideas are sometimes not really welcome, but rather "hold up the business". The inventor might hear something like: *"Another new idea, Mr. Smith? You must have a lot of time on your hands!"* Or: *"What did you make up this time, Mrs. Spencer? Has the order for part X actually been completed?"*

What can be done instead? Creativity of people is a prerequisite, that cannot be forced, only encouraged. In our experience, other prerequisites for innovation are above

© Springer-Verlag GmbH Germany, part of Springer Nature 2020
M. Jantzer et al., *The Art of Engineering Leadership*,
https://doi.org/10.1007/978-3-662-60384-0_7

all expertise, curiosity and a certain pressure to find solutions. When technical contradictions have been identified creativity methods support finding solutions.

Innovation requires leadership to create an environment that encourages new ideas and their subsequent development. Ideally, this process brings product engineers, customers, users and decision-makers together. This should not happen by chance.

So what is the leadership task in innovation? Engineering organizations are often designed to ensure engineering quality in order to minimize the technical risk of products launches. But innovation is always a risky journey into the unknown, People who take risks for innovation will run into conflicts with an organization focused on risk reduction. In this conflict-prone field, leaders ideally act as "promoters" or "owners" of innovation topics. At important milestones, they provide orientation and make the necessary decisions to remove impediments.

In order to activate creativity, leaders demand solutions to challenging problems. These can appear to be virtually "unsolvable", such as "reduce the package size by half" or "use only off-the-shelf parts". Difficult challenges encourage engineers to think outside the box.

## 7.2    Thinking Outside the Box

"Think outside the box!" is easily said. In order to explain how "outside the box" relates to innovation and what leaders can do to enable it, we take a look at how humans learn.

During learning, our brain makes new neural connections. By repeating and practicing, the new neural connections are activated repeatedly and reinforced. At some point, what has been learned is applied largely automatically, i.e. as soon as a trigger is activated, a routine starts. This saves energy and we can focus our attention on other things.

Much of what we think and do is firmly anchored in the neural structure of our brain without us being aware of it. We can drive a car on a familiar route, while thinking of something completely different. At the end of the journey, we may not remember its details.

This also applies to everyday engineering. For example, product engineers think automatically in the existing technology because it is practiced. This is beneficial to speed-up incremental development. However, something new can only arise when new connections form in our neural structures. To achieve this, we need new impetus. In daily project work, driven by deadlines and efficiency, new impulses are easily put aside. Leaders can do a lot to prevent this:

- Attendance at expert conferences, exchanges in expert networks (internal and external), customer visits, supplier visits, employee job rotation, diversity in teams, etc. are suitable for promoting lateral thinking and exchange, if they are focused on a specific task.
- Formulate tasks to break with the usual way of thinking (e.g. contradictions, paradox tasks).
- Participate in innovation topics for projects other than your own.

- Provide opportunities and time to try out solutions and carry out experiments. This promotes learning and leads to new ideas.
- Continuously analyze competitor products to explore alternative solution strategies (this also counteracts the "not-invented-here syndrome").
- Systematically combine information from many sources to provide an overview of the respective technology development.

## 7.3   User Needs

Innovation requires knowledge of customer and user needs. The following concepts support a systematic approach:

- Design Thinking [1] identifies user needs, develops prototype solutions and records user experiences at an early stage. After evaluation, they are transferred to the next prototype. In Design Thinking, a design is never regarded as "finished", more and more mature prototypes are created in a continuous learning process.
- User Experience or UX [2] is an approach to user centered thinking. UX describes all experiences that users have when interacting with the product. In the so-called Customer Journey, all user contacts with the product are recorded in order to document as precisely as possible to what extent it satisfies the respective needs.
- The Lean Startup approach [3] transfers "learning by doing" to the company foundation itself. Products or services are launched in the market at an early stage in order to verify value hypotheses. User feedback is obtained at an ongoing basis and provides information on customer wishes and market needs. The result is a value and growth hypothesis on which the start-up is based.

---

**Example**

We illustrate this with the example of the e-bike: Riding a bike makes you mobile, provides sporting activity and is simply fun. For example, customers ride their bikes to work. They want to cover longer distances and drive effortlessly uphill. Others want to be able to keep up with the group on longer leisure tours. E-bikes are a solution for all of this. Now, however, mobility range is becoming one of the central customer needs.

The electric drive of an e-bike, or pedelec, requires a control unit, electric motor and battery. The range depends on the size and weight of the battery. For a long time, heavy and block-like batteries hanging on bicycles were a deterrent for cyclists. Lithium-ion technology was able to meet the customer's need for range at an acceptable weight and attractive size. That is why e-bikes have been conquering the bicycle market since around 2005.

The history of e-bikes at Bosch is based, among other things, on thinking outside the box: existing technologies from power tools and automotive markets were applied. Many employees were passionate cyclists themselves and were able to formulate their own needs. Workplace cyclists, recreational riders, sporty mountain bikers, mixed teams of men and women, created a comprehensive picture. For example, employees could use prototypes in their free time and for short business trips.

The executive leaders themselves were also enthusiastic about cycling. They quickly realized that the market and product ecosystem (OEM, frame manufacturers, suppliers, marketing and sales, wholesale and retail, etc.) were completely different from power tools and the automotive business. Within these existing business divisions, the further development of innovation was therefore hardly feasible.

That is why Bosch founded its own business unit in 2009 based on lean startup principles. This created the right framework conditions to make innovation possible. After a short development period, the first generation of the e-bike was launched in 2011. Bosch E-Bike Systems has been the market leader in Germany and Europe since 2012.

With the founding of the startup Bosch E-Bike Systems, the small team also moved into its own building. Initially, this was an unusual picture at Bosch: The office seemed to be a laboratory. Parts and partly assembled bicycles were everywhere in the corridors and on the desks. The employees came from the most diverse areas, but one thing was quite clear: This team worked on a common vision.

The ideas for the e-bike were initially developed in the current business areas of Power Tools, Electric Drives and Automotive Electronics. Leaders in these areas have facilitated and motivated ideas and promoted exchange. They have enthusiastically promoted their ideas and collected the resources to develop them further. They saw the opportunities, and took high risks. They believed in their product, convinced stakeholders and fought for it.

## 7.4    Creating the Framework for Innovation

People from different perspectives contribute their knowledge to realize innovations. The results are often not attributable to individuals, rather they are the result of a team effort. Not everyone can cope with this. It may be even in contrast to the organizational reward system, which is focused on individuals. Leaders can start changing this reward system and following the principle of "team success first, self-interest second". Glorification of individual inventors is one of the enemies of innovation. Innovation requires shelter from efficiency pressure in daily project work, e.g. dedicated innovation teams for a certain period. Nevertheless, a specific task with a goal and a deadline should be formulated. Established budgets are needed to advance ideas, even if they do not turn into innovation in the end. Usually you can learn a lot from the result.

What about the diversity in your team? We tend to surround ourselves with people of the same views and ways of thinking. This makes sense for recurring tasks, where similar perspectives and understandings can help to avoid negotiating every detail. However, "diversity instead of simplicity" applies to innovations. You shape diversity, e.g. through your selection of people: other cultures and domains, women and men, age group, or through exchange programs.

## 7.5   Leading the Decision-Making Process

Suppose you have created favorable conditions and your team has many good ideas within a particular field. Now it is about selecting the most promising. At the end, there will be one solution, The selection is the result of a process, which has to take opportunities and threats into account and weigh them up.

Leaders and teams constantly make decisions in this selection process. Solutions must be sufficiently elaborated to enable decisions. Often combinations of different concepts are effective. This selection process is hard work, since different solutions must be evaluated every step of the way. Good solutions must be distinguished from those that have insufficient potential for success. Decision-makers have to prioritize the value contribution for users.

So far we have reported on major innovations, on sources for innovation, boundary conditions and the process of innovation. Even if an idea does not turn into a completely new product, the same principles are valid in more evolutionary innovations, e.g. in optimizing manufacturing processes or engineering methods. Ideas and possible innovations can be found in daily work and from every employee.

Perhaps you are wondering what you personally have to do with innovation: courage, curiosity, learning ability, willingness to take risks, and determination create innovations. Ultimately, it is you as a person, your believes and your picture of the future that create and promote innovation. This often means that you have to defend ideas against resistance, convince sceptics, and prove your perseverance, see Fig. 7.1.

**Fig. 7.1**  Tasks in leading innovation

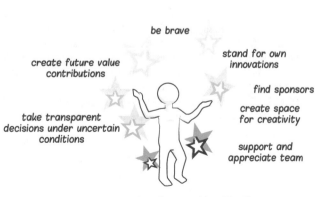

At the end of this chapter, we do not want to conceal the fact that luck is always part of innovation. As a leader, however, you have a certain amount of control whether new ideas are born continuously and whether these ideas get the chance to establish themselves successfully in the market.

▶ **Practical Tips**
- Observe how your team develops ideas and how you react to new things. Implement an innovation roadmap for each team. Do not just focus on functional improvements in your innovation work.
- Ensure that hypotheses are validated early on the market.
- Talk appreciatively about stopped innovation projects and record what you have learned.
- Formulate tasks for your team that lead them beyond their accustomed ways of thinking
- Work out contradictions in terms of content and focus your team on their resolution.

**The Most Important in Brief**

Innovative strength is measured exclusively by market success, which is achieved only through convincingly satisfying the needs of customers and users.

In order to create something new, established paths must be left behind. Courage, curiosity, willingness to take risks, and determination are of central importance.

Innovation is often difficult knowledge-based work and requires dealing with conflicts.

# References

1. Hilbrecht, H., Kempkens, O.: Design Thinking im Unternehmen – Herausforderung mit Mehrwert. In: Keuper, F., Hamidian, K., Verwaayen, E., Kalinowski, T., Kraijo, C. (Eds.). Digitalisierung und Innovation, S. 347–364. Springer Fachmedien, Wiesbaden (2012)
2. ISO 9241-210: Ergonomics of human system interaction—Part 210: Human-centered design for interactive systems. ISO, Genf (2010)
3. Ries, E.: The Lean Startup: How Today's Entrepreneurs Use Continuous Innovation to Create Radically Successful Businesses. Penguin, London (2011)

# Requirements Engineering

**8**

▶ This chapter shows success factors for excellent requirements engineering. Expectations are formulated, methods are presented and favorable leadership behavior identified.

Requirements engineering is probably the most important part of the development of new products. It starts at a very early stage and is the basis for the later success of a new product. At the same time, it is the development step with the greatest uncertainty. Early in a project, requirements are still unclear and different technical concepts are competing with each other. Often the various stakeholders have conflicting or contradictory requirements. This leads inherently to insecurity and conflicts. A colleague of ours explains good requirements engineering with the following analogy:

---

**Example**

"This is how vacation planning has worked in my family for a few years: In January, my daughter asks where we will go for summer vacation and suggests a long-distance trip. My son wants to go to the sea, preferably without a long journey. My wife wants to rest and enjoy culture. I'd like to go hiking in the mountains. It is impossible or very expensive to reconcile all these requirements. If one person wins, the rest is in a bad mood. If we try to persuade each other, we'll have a fight. We've already tried all that. Meanwhile we do it like this: Everyone says what is important to them. We create some concepts, which consider the needs for sports, culture, relaxing etc. We build these concepts together with pictures and plans. In this way we come to a joint decision. The results are sometimes surprising, but everyone is looking forward to the vacation."

© Springer-Verlag GmbH Germany, part of Springer Nature 2020
M. Jantzer et al., *The Art of Engineering Leadership*,
https://doi.org/10.1007/978-3-662-60384-0_8

Actually, we could close the chapter with this story, since it contains all the essential elements of successful requirements engineering: All stakeholders win in the end. The process must be well managed for it to work. All parties involved are committed to find a win-win solution. A planned approach facilitates the resolution of conflicts. A good solution does not have to meet the original requirements of the stakeholders. They may be inspired by the needs and ideas of others and open their minds for alternative solutions. Pictures and prototypes help, because they are inspiring.

However, since requirements engineering is the biggest single success factor for projects, we would like to take a detailed look at how to lead it.

## 8.1   Objective and Expectation on Requirements Engineering

At the beginning of product development all participants have a common goal: develop a product that will be successful on the market and that will help our company to succeed.

In detailing this objective, there are often very different ideas about what should be developed. Requirements engineering is necessary to bring these ideas together. It is therefore not productive if product managers hand over a finished requirement specification booklet to the development department stating "This is the only way we will survive in the market!" Neither is it effective for development or production managers to try to impose a dictate of feasibility stating "For technical reasons this is the only way!" Only together they can specify a concept that is both marketable and feasible. As shown in Fig. 8.1, this is a recurring task throughout the entire development.

As in the example of our colleague, this means that it does not matter whether each party realizes its original idea. Requirements engineering is excellent when all stakeholders are winning [1], e.g. they are convinced that the product will help them achieve their goals.

The early phase of a project is often characterized by uncertainty, changes of direction, and conflicts. The ideas of project participants are very different. There is a great deal of uncertainty about what is really marketable and feasible. Conflicts arise from this. Conflicts are productive, if the participants work out a common vision of the product based on the available knowledge. In unproductive conflicts, however, content is disputed that could be clarified objectively and beyond doubt. For example, it makes sense to discuss a possible extensibility of an architecture for future requirements, while it does not make sense to argue about the fulfilment of an environmental requirement that is measurable. We can therefore formulate the following expectations:

- Requirements are formulated as far as possible based on figures, data and facts to prevent avoidable uncertainties.
- Conflicts arising from different goals and ideas are clarified and all stakeholders look for win-win concepts together.

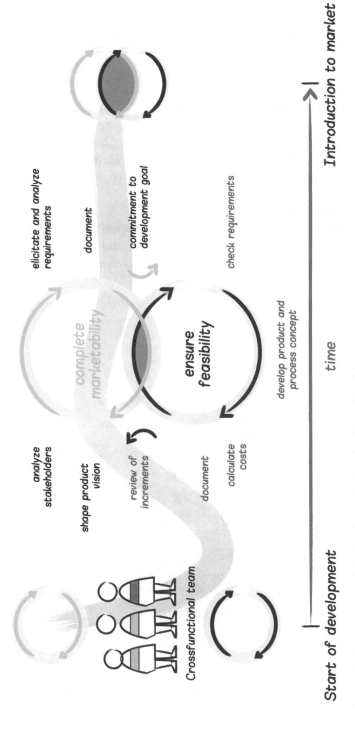

**Fig. 8.1** Projects are successful, when they are both marketable and feasible

- In order to resolve conflicts, managers ensure good and transparent decisions. They reduce uncertainty and establish the team's ability to work.
- In order to achieve transparency, the team uses suitable procedures and state of the art requirements engineering tools.

## 8.2    Approaches and Success Factors

Teams collect, test, evaluate and document requirements. They work according to the state of the art, as described for example in Rupp's book [2]. What are the contributions of the engineering leaders?

**Stakeholder Management**
To turn all stakeholders into winners, you must first identify them and then determine with whom you must work intensively. A "stakeholder map" is suitable for this, see Fig. 8.2. It can be used to develop a suitable strategy for communication and cooperation with stakeholders.

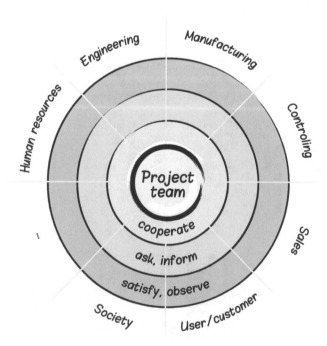

**Fig. 8.2**  Identification of stakeholders and derivation of a communication strategy

Since the cooperation with the stakeholders needs to function throughout the entire project duration, it is essential to find a common ground for win-win-solutions, see also [3].

**Elicit Requirements**

Leaders pay attention to the application of suitable methods and the selection of sources for requirements engineering.

A variety of methods have been established for elicitation (QFD, UX, Design Thinking, Pretotyping…). If you reach an early agreement with your stakeholders on approaches and sources, the resulting requirements will be accepted. It is essential to use lessons learned from previous projects and competitor analyzes as sources.

**Clarify, Evaluate and Prioritize Requirements**

Once teams have collected requirements in the form of user stories or specifications, the result is often a picture that corresponds to the vacation wishes of the colleague's family. The ideas of different stakeholders differ considerably. At this point, you can help your teams by commissioning concepts that are then discussed with the stakeholders. Step by step the specification for a concept that satisfies all stakeholders is created. For this interaction between requirements analysis and concept development, the joint consideration of functional and solution structure, as shown in Fig. 8.3, is helpful. The hierarchy of requirements naturally becomes visible. Good structure and hierarchy makes it easier to master the multitude of requirements.

In the requirements review, the function and solution structure are good tools for determining at which level decisions have been made and which level is currently considered in the review. Reviews ensure that requirements engineering has the right priority for your employees. In the review you improve the content and structure of the requirements together with the team. Special emphasis is to be placed on the following properties:

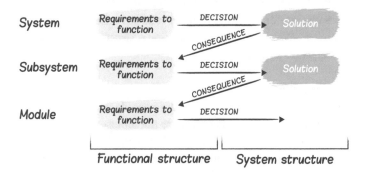

**Fig. 8.3**  Derivation of requirements and solutions in a hierarchical system model

- Correctness: Will the stakeholders accept the requirements?
- Consistency: Do requirements contradict each other?
- Clear: Is there no room for interpretation?
- Completeness: Are all stakeholder needs considered?
- Traceability: Can requirements be traced back to a stakeholder either directly or through inheritance?
- Verifiability: Can fulfilment be verified later?
- Evaluability/priority: Is the requirement small enough that it can be evaluated in terms of its importance and be prioritized?

In the best case, requirements are also available as executable models, diagrams and graphical visualizations. Then requirements reviews lead to exciting discussions.

**Document Requirements**

The documentation of requirements is not an activity that is carried out at the end of a requirements phase of a project in order to be able to file a fixed specification book- let. New requirements are formulated throughout the entire life cycle of a product. The product is maintained, new features are added or the product is adapted to new stand- ards. These changes have to be documented so that they can be discussed and decided. However, it often happens that teams document requirements only if they are essential for outsourcing. Typically these teams use office programs for documentation of require- ments. Only a few requirements are filed and much remains in the minds of the develop- ers. Other teams maintain and update well-structured requirements documentation for all components. They manage requirements using professional software. They discuss new requirements with stakeholders systematically.

It is the task of leaders to provide their teams good tools for requirements engineer- ing and to discuss the requirements documentation regularly. Depending on the degree of novelty of a development, teams spend up to a quarter of the total project duration in the requirement and concept phase. In our consulting practice, we repeatedly find situa- tions showing that this phase is organized inefficiently. Either this phase takes too long and leads to demotivation of the team, or the results are unsatisfactory and endanger the success of the project. With a suitable organizational structure, leaders can support their teams in using this phase productively.

**Composition of the Project Team**

While forming a project team, it can be difficult to decide who should be in the team and who should support the team. If stakeholders are not represented by team members, spe- cial emphasis has to be put on implementing their requirements.

Depending on the project task, it may make sense for purchasing and production to be an integral part of the team (so-called simultaneous engineering teams). Sales and marketing are often not part of the team, but are involved in the work as stakeholders.

The intensity of the required collaboration is a good indicator to balance between agility and interface optimization. Since changes of direction with large teams are difficult and slow, smaller teams are preferable.

**Competence Development**
At the beginning of a project it is necessary to bring the competence level on requirements engineering of the team to a common level and clarify with them which results and procedures they are expected to follow. For this purpose, a trained requirements engineer should be part of the team or at least be available as advisor.

**Decision and Escalation Model**
To work well and quickly, it is helpful to determine who decides which requirements. Typically, the project mentor or agile entrepreneur decides about:

- User Stories to cover
- Operating conditions to be covered
- Quality targets
- Targets for product costs and investment budget.

Project managers or product owners ensure that central decisions are made, e.g. on:

- Product concept and architecture drivers
- Division of tasks and responsibilities within the team
- Procedure and demands on the procedure.

In matrix organizations it is necessary to clarify the tasks, rights and responsibilities between project, line organization and systems architect to be able to resolve conflicts quickly. All roles have to be aligned explicitly in the early project phase.

**Requirements Management**
Effective handling of requirements changes is an important key to the success of a project. Changes ensure marketability and feasibility in a changing environment. However, if the rate of change exceeds a value of about 1% per month, teams may lose their rhythm. To prevent this, it makes sense to form a committee that accepts, evaluates and prioritizes new requirements or changes. This change control board accompanies a product development through its entire life cycle.

No matter whether you are planning a small change to an old product or want to develop a complex new product, the following generally applies to requirements: garbage in—garbage out. Poor requirements lead to poor results.

▶ **Practical Tips**

- Qualify yourself in requirements engineering and formulate your expectations of the team.
- Reflect on the way you work: Is requirements engineering structured hierarchically? Is communication with your stakeholders well organized? Are requirements consistently documented?
- Before discussing solutions, make sure that the requirements are clarified independently of the solution and that acceptance criteria are defined.

**The Most Important in Brief**

Requirements engineering is the most important single process in product engineering.

Requirements engineering is excellent when we turn all stakeholders into winners. The continuous involvement of stakeholders in the design process is an important success factor.

A systematic approach and the use of powerful documentation tools are indispensable for excellent requirements engineering and the basis for quick decisions about conflicts.

# References

1. Boehm, B., Lane, J.A., Koolmanojwong, S., Turner, R.: The Incremental Commitment Spiral Model, p. 2. Addion-Wesley, Pearson, New Jersey (2014)
2. Rupp C, die Sophisten: Requirements-Engineering und -Management: Aus der Praxis von klassisch bis agil. Hanser, München (2014)
3. Nasa SP-2007-6105 Nasa Systems Engineering Handbook, p. 35. NASA, Washington D.C. (2007)

# Design the Product Architecture

9

▶ This chapter explains the essential characteristics of architectures using the structures of cities as an example and relates them to product architectures. The limits of decentralized decisions are shown. Architecture drivers are identified and the necessity of conscious architectural design is explained.

The second part deals with leadership aspects, competencies and basic attitudes in architectural design.

## 9.1 Create Architecture

Let us take a little detour to reflect the difference between evolving and planned architectures. Looking at the maps of two cities [1] the difference between grown architecture (Algiers) and planned architecture (Cologne) becomes apparent (Fig. 9.1). Both have their advantages and of course their limitations.

Evolving structures follow local needs spontaneously and consider local conditions. As long as the requirements for further development remain locally limited, evolving structures are good. They have their advantages because they implement local requirements fast and good. However, more extensive changes are difficult to implement in these architectures. If, for example, the traffic in Algiers would require more flow, deeper intervention in Algiers' road network would be necessary and many structural elements (houses, infrastructure,…) would be affected at the same time. A "re-engineering" would be time-consuming and expensive.

Planned cities start with a regulatory framework that includes a functional structure and a supply and disposal network. This is most evident in the road map.

Cologne was once founded by the Romans and followed an architecture that had already been explored and tested many times. The original structure is still visible today.

© Springer-Verlag GmbH Germany, part of Springer Nature 2020
M. Jantzer et al., *The Art of Engineering Leadership*,
https://doi.org/10.1007/978-3-662-60384-0_9

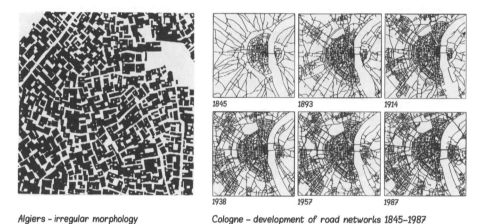

Algiers – irregular morphology          Cologne – development of road networks 1845–1987

**Fig. 9.1**  Structures and planning of cities

The subsequent development of the city followed the structures already in place. They were extended and further developed. The original layout remained sustainable for many years and investments could be used for a long time. In such a case, evolving needs can be continuously implemented within the existing structures. A thorough re-engineering is not necessary. Development can be evolutionary. This is cost-effective over time and relatively fast and flexible. Despite its long life cycle, the city remained modern and attractive.

Well thought-out urban planning pays off in the long term. It can react to changing needs. However, it also limits the ability to make local decisions. Decisions must always keep "the big picture" in mind. Not all local needs will be met since decisions are made from a higher-level system view.

Let's move back from this little excursion about urban development to product development.

If you are responsible for a simple product with limited lifespan, think carefully about how much you need to invest in robust product architecture.

If you are responsible for a complex product that you plan to establish on the market for a long time you should consider investing in a robust product architecture. The more complex the product is, the more the initially high costs of architectural work will pay off over the product life cycle.

The prerequisite is the robustness of the architecture. In architecture design, the term robust is used if volatile requirements do not result in significant changes to the product architecture. It may be necessary to replace or add individual elements, but the adjustments can be easily implemented through individual elements.

Product architecture: What is that? It is not so easy to grasp, because architecture is above all an intellectual work. It is a regulatory framework. This framework comprises on the one hand design principles (including perceptible aesthetics), detailed design rules

Fig. 9.2   The BOSCH ESP-system

and methodical approaches. On the other hand, architecture introduces logical and physical structures [2, 3].

Let us look at the architecture of a technical system as an example. We dissect an ESP® unit for this purpose (Fig. 9.2). The external physical architecture is immediately visible: a connector, an electronic module, a hydraulic module, an electric motor. The internal physical structure would become visible if we had disassembled the device.

The logical structure is less obvious. To understand how it was created, we start with the system requirements.

First, the main requirements of all relevant stakeholders are compiled: The product architect gathers the customer requirements (see Sect. 8), but also those of all internal stakeholders as well as general market constraints. This includes the foreseeable changes over the product life cycle (Fig. 9.3). Potential changes like new laws or new customer wishes are taken into account based on scenarios.

Now you can develop a logical structure within the requirements. Since this is a complex process, it will rarely succeed right away. Once you have created it in all dimensions

Fig. 9.3   Dimensions of requirements

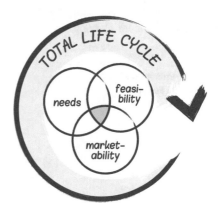

you can assess whether the architecture is good. This cannot be done a priori, but only posteriori. Therefore an iterative and incremental approach is recommended, starting at the highest system level that is accessible. This system level should be at least one level above the offered product or service. From there you work successively through the system levels ("from rough to detail"). At each level, several alternative architectures and proposed solutions should be considered and deliberately selected (Fig. 9.4).

Different and clearly separated views (see Fig. 6.5 and Table 6.1) help to divide the whole problem into understandable pieces and make it discussable ("separation of concerns"). A view describes one of the desired characteristics of the entire design problem. Characteristics can be, for example, reliability, manufacturability, costs, or other essential quality attributes. Using the example of the city of Cologne, one could look separately at the drainage of water. Whether the proposed solution is solving the overall problem can only be seen when the partial solution for the water drainage is integrated into the overall architecture. Now one can assess whether the overall problem—urban planning—has been satisfactorily solved.

If an alternative has been chosen at the respective system level, we recommend a reflection on the robustness of the architecture at least at the next higher system level ("Continuous Integration").

In our opinion, the development of architecture requirements and their solution follow a logical sequence, but they cannot be separated in time. The discussion if requirements can be fulfilled within the framework of a desired business model will go hand in hand with architecture and solution definition. Thus logical structure and the design rules applicable to this product (see Fig. 6.1 and 6.2) are created simultaneously.

The example of the ESP® system shows that this requires knowledge from the following domains: electrics and electronics, mechanics, production process development, control engineering and computer science, but also special disciplines such as noise, electromagnetics and much more. At the same time, a sound knowledge of the market, customers, potential partners and suppliers as well as possible business models is required.

For complex products and services the domain expertise can hardly be found in one person. A cross-functional and interdisciplinary team and an architect are needed. The architect discusses opportunities and limitations with all domains (Fig. 9.5). Above all he guides the team to combine the opportunities of individual domains into surprising

**Fig. 9.4** Principles of architecture design

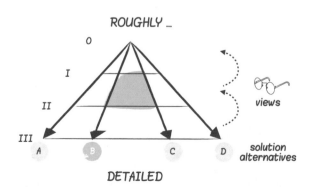

solutions. This team is not necessarily constant, because the participants for system level 3 can be different from those at system level 0. The architect has to work with many different people who represent different interests.

He also has the task of identifying architecture drivers, creating technical and logical structures and defining the relevant design principles and rules.

What are architecture drivers? Architecture drivers are requirements that decisively determine the product.

Let us get back to our ESP® example. Let us look at a few requirements on the upper system level:

- It has to work, of course. This means that the longitudinal and transverse guidance of the vehicle is supported by wheel-specific braking interventions.
- It should be cost effective
- It should be easy to integrate into the vehicle.
- It has to be reliable.

What else does the market need?

- The car market is characterized by increasing diversification and individualization. This leads to an increase in the number of vehicle projects and even more equipment variants.
- ESP® is standard in many markets. It is therefore installed in both very small and very large vehicles which have very different braking systems.
- The market is striving for autonomous driving cars, which will need additional braking functions. It is unclear when they will be needed.

**Fig. 9.5**  Architecture team

COLLECTIVE IQ

From this rough and incomplete point of view, there are four drivers for the system architecture:

- Functional and product characteristic costs
- Easy integration
- Functional scalability
- Dependability

For this reason, the architecture of ESP® systems follows a modular strategy. The basic architecture is kept constant and individualization is realized using configurable module kits—to a limited extent in hardware and to a greater and increasing extent in software.

The work of the architect is not apparent in the final product. However, a good architecture decides about flexibility and thus about long-term marketability in a volatile environment.

The ESP® development is a complex task where many people work together for a considerable duration. Therefore, the performance of the architecture team can usually only be assessed in retrospect. Nevertheless, there is ample evidence [4] that investing in a sophisticated architecture will certainly pay off for complicated and complex projects.

How to lead in such an environment? How does an architect lead the team? And how do you lead your architects? After all, this step is crucial for the sustainability of development efforts and thus for their effectiveness and efficiency.

## 9.2    Leading the Architecture Design

Architecture design creates structures in a complex environment. Architecture design is not only about designing the system to be developed. The product development system and the people system have to be designed as well. Their cooperation, decision-making processes and design guidelines for the product or service are organized. The design of the system architecture is a technical leadership task in itself.

Let's take a look at the role of the architect. He is a central person because, as we have explained, decisions should always be taken from a systemic point of view. Thus, important product decisions are always made by the architect. At the same time, the architect depends on open cooperation with all domains involved. Developing an architecture by mandate is not the best idea.

At the beginning of the product life cycle the system has to be designed for the first time. Market opportunities and options should be explored. It is not so much about making decisions yet, but about working out opportunities, about breadth and thinking big.

Put yourself in the position of the architect. You are looking for opportunities in this phase with a cross-functionally staffed team. Leadership always starts with oneself. So, in the development phase you lead by adopting an open and curious attitude. This helps you to be inspired by individual contributions, to explore perspectives and needs of other domains and to grasp new opportunities. Curiosity opens up the own mind. In addition, the curiosity of the leader will unlock the team's curiosity. The architect accepts contributions and develops them further. In this phase ideas should not be answered with a "BUT, what about…", rather than with a "interesting, AND…".

At the same time, however, the architect must not get lost in details. He must channel the discussions sensitively and focus them on the target. We always see both movements in successful teams: opening up for new solutions and refocusing. Successful teams do not see this as a contradiction, but as part of learning.

Curiosity supports the understanding of other people, to get to know and appreciate their way of thinking, to accept different approaches and to open themselves empathetically. Especially in complex design tasks, those who appreciate a variety of approaches are more successful. To work out an approach that is feasible and promising for everyone and to adopt it without prejudice is good advice for the architect. Complexity demands flexibility in thinking and acting, and with a curious attitude this is easy to achieve.

Successful architects are strong personalities, but they do not act as "strong" doers. They create the right solutions from many contributions. This is a subtle way to lead, based on empathy and creating sympathy. It is a very decisive characteristic. Because a diverse team makes the most interesting contributions, they are potentially conflicting. Innovations arise from conflicts, from contradictory goals. It is exactly in this situation that the architect is dependent on people opening up to contribute their real view. In our experience, this succeeds when mutual understanding, curiosity and interest for each other and the interests of others in the team go without saying. At the same time, people need orientation. Successful leadership of architectural design creates openness on the one hand and on the other hand continuously integrates it into a target-oriented solution. This invites everyone on a journey, which is only just taking shape through their actions.

To be able to integrate everyone into the team, you need, among other things, a solid expertise. The architect should be able to understand where the journey is going and what needs to be done along the way. Architects will probably not have this competence in all domains, but with curiosity they will understand all domains to such an extent that they can take up their relevant constraints, opportunities and patterns.

Are we looking for a superwoman/man? Yes and no. Such leaders do not appear from nowhere, but both technical and interpersonal skills can be learned. With a curious and solution-oriented attitude this will succeed. In our experience this means a few years of apprenticeship (compare also [5]).

Returning to a familiar theme: another important attribute is the trustworthiness of the architect and the fundamental trust in team members. This is expressed on the one hand in integrity (Sect. 2.2), and on the other hand in differentiated appreciation—for good contributions as well as for those that do not lead to good results.

If you are trustworthy, people will honestly talk about the opportunities of their domain to solve a problem—even if they have not caused the problem and/or take on a burden to solve it. Trustworthiness simply opens the solution space and makes leaders—and architects—more successful. You need trust not only for your team, but also for your stakeholders and colleagues. The balancing act is not always easy. However, we can see that integrity is an essential key to this. And integrity can be supported by clarity about oneself, one's personal motivations, and one's firm inner attitude. This is one of the reasons why self-leadership is so important.

Trust can be built in daily work through open and empathic communication. We do not mean pain avoidance, but the empathic discussion of sensitive topics with the aim of creating clarity. At the same time, we have experienced lasting success by "deep" and appreciative discussion. Development is ultimately about resolving technical conflicts, but there are people behind the topics. Therefore, we always have an interpersonal dimension in the technical conflicts (see Sect. 20.1).

This brings us to another aspect of leadership in architecture design: clarity of goals and roles. What is my contribution and what do I expect from others (see Sect. 18.2)? Every day we experience that all sorts of implicit assumptions are made. People implicitly assume all sorts of things that are not necessarily shared by others. It is therefore beneficial to readjust roles, goals and expectations throughout the project.

The resolution of content-related conflicts is part of a solution-oriented learning culture. Learning involves taking knowledge from others, discussing it candidly and translating it into solutions. The architects will always learn something they did not know before. This is inherent in architecture design.

Once the architecture is developed, the structure is in place and the design rules have been agreed, the architect must enforce them. Unjustified deviations from the agreements are to be prevented to ensure success. In this phase we might refer to the enforcement mandate, of course with an empathic attitude, because the trust gained is still needed.

We have sketched a picture of an architect as a mature and strong personality. You lead the architect in the same way as the architect leads his team: with ADIOS (Fig. 2.3) and empathy.

Finally: How do you recognize a good architecture? How do you know when the architecture is done? It is the elegance of the solution: everything makes sense and is coherent and somehow "simple". This applies to the city map of Cologne as well as to the modular design kit of an ESP®.

▶  **Practical Tips**
- Make yourself aware of the architecture drivers of your solutions retrospectively and determine the limits of your architecture. This allows to evaluate change requests transparently.
- Build up competencies and experience in architecture design in your teams.
- Work out a role model for the architect.

| The most important in brief |
| --- |

Robust architectures allow an efficient implementation of volatile requirements. Despite the high initial investment, this approach leads to economically superior solutions in the long term. The most important architecture principle here is the "separation of concerns".

Architecture drivers are requirements that decisively determine the product.

Architects must integrate many disciplines and understand their opportunities and challenges. That is why openness and curiosity are favorable basic attitudes.

# References

1. Curdes, G.: Stadtstruktur und Stadtentwicklung. Kohlhammer, Stuttgart (1993)
2. Maier, M. W., Rechtin, E.: The Art of Systems Architecting. CRC Press, Boca Raton (2009)
3. ISO/IEC/IEEE 42010:2011(E): Systems and software engineering – Architecture description, Geneva (2011)
4. Eric, C. Honour: Systems engineering return on investment. PhD-thesis, University of South Australia, Adelaide (2013)
5. Godfrey, P.: Building a technical leadership model, 26th Annual INCOSE International Symposium (2016)

# Model Based Engineering

# 10

> All engineering work is based on models. Models describe the products we develop and their interaction with their environment. In this chapter, we describe the concept of model-based development and its value proposition. We discuss the implementation of model-based development and the expectations that the engineering leader communicates.

**Example**

After a heavy storm hit the mediterranean island of Antikythera in the spring of 1900, Greek sponge divers discovered sculptures and amphorae on the sea floor. Over the next few years, researchers found the remains of a large merchant ship that sank in a storm around 70 AD. The most enigmatic find is the remains of a complicated work of gears, hands and scales (Fig. 10.1). For over one hundred years, historians have been trying to figure out how this device from the 2nd century BC could have worked. In the early 2000s an international team of researchers successfully decoded the inner workings of the more than 30 gears: The Antikythera mechanism once was a mechanical, perpetual calendar for predicting important future events.

In ancient times, the positions of the moon and the planets, as well as special constellations of the stars, played an important role in public life. Public events and celebrations were scheduled according to the position of the moon. Knowing the exact date of a lunar eclipse determined the outcome of a war. Predicting the position of celestial bodies and the date and time of lunar and solar eclipses was of great practical and strategic importance. The Antikythera mechanism could display all this on its scales. It is not only an astonishing example of ancient astronomy and mathematics, but also a proof of excellent model-based product development. Based on observations about celestial events such as lunar eclipses, ancient mathematicians

© Springer-Verlag GmbH Germany, part of Springer Nature 2020
M. Jantzer et al., *The Art of Engineering Leadership*,
https://doi.org/10.1007/978-3-662-60384-0_10

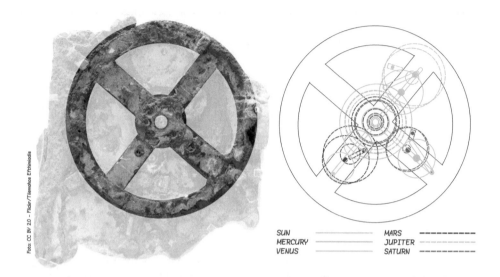

| SUN     | ———————————— | MARS    | ------------ |
| MERCURY | ———————————— | JUPITER | ------------ |
| VENUS   | ———————————— | SATURN  | ------------ |

**Fig. 10.1**   Artifact of the Antikythera mechanism, an ancient Greek mechanical computer

developed models to describe the elliptical orbit of the moon. The developer of the Antikythera mechanism used these models and implemented them in a technical solution using a hand-operated gearbox.

In the same way the ancient engineer converted mathematical models into a gearbox, developers today make technological advances based on understood cause-effect relationships represented in models. Product properties are modeled, optimized and afterwards realized. As long as there is no realization e.g. in the form of prototypes, the modification costs are low and the solution space is large. This offers a good chance of finding cost-effective solutions for the desired product properties. In an approach where modeling plays little or no significant role, optimization takes place only after implementation and test. The solution space is typically limited and optimization is costly. (see figure Fig. 10.2).

The focus on modeling is accompanied by a shift in the development effort into the early stages of development (frontloading).

## 10.1   Value Contribution of a Model

Models are of value for product engineers if they can base their design decisions on them and if test results can be reliably predicted. The purpose of testing is to verify the models. Models that are not used because no one trusts them are worthless. Simulation services that do not make a significant contribution to the design add no value and are therefore wasteful.

Fig. 10.2 The cost to
achieve a desired level of
quality increases over the
product lifecycle

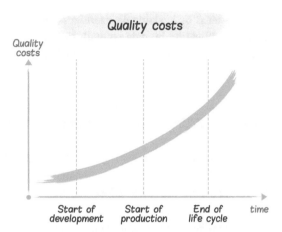

Developers strive to develop fast, make accurate statements, and remain flexible to changes in requirements (see Fig. 10.3). Models are valuable to them, if they enable them to achieve these objectives.

In practice, this means that the value of models lies in accelerating design and verification, or in the reduction of realization risks.

A model is excellent, if its value contribution to fast, agile and accurate engineering can be achieved with minimal effort. In addition, it should extend the engineer's competence. This means that modeling is not about completeness or intellectual brilliance, but only about immediate and long-term usefulness.

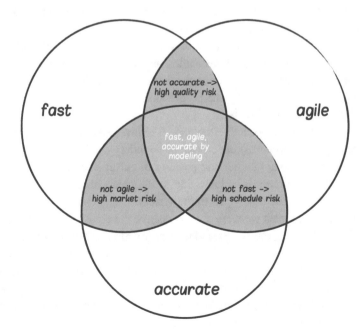

Fig. 10.3 Objectives in engineering are to be fast, agile and accurate

Long-term or sustainable usefulness means that a model can be used outside the current application boundaries with little effort. Among other things, models therefore require a solid scientific basis and a clear description of the modeling depth, i.e. the described and neglected effects, as well as the premises on which the model is based.

## 10.2  Leadership Expectations on Model-Based Engineering

An essential prerequisite for the implementation of a consistent model-based engineering is the clear expectation of the engineering leaders to follow a methodical approach. From our experience, these expectations are rather self-evident for areas with high R&D budgets and hard to reach for areas with minimal R&D budgets. It is often overlooked that model-based engineering can be advantageous even with the smallest development budgets. Differences in modeling depth and use of resources do not mean that the general expectation for procedures and results has to be lowered. The following example of one of the authors shows how a considerable value contribution can be achieved through modeling, even with a limited resource.

### Example

In the development of a self-priming dosing pump for aqueous media, one focus of the design was on the robustness of the pump when the fluid in the pump froze. In order to achieve robustness against freezing, the pump had to withstand the increase in volume during freezing without damage. It had to be sufficiently elastic. For delivering a constant flow rate against variable back-pressure, the pump had to be as rigid as possible. This apparent contradiction had to be resolved.

We formed an interdisciplinary team of one designer, one simulation engineer each for structural mechanics and fluid mechanics, and one systems engineer.

We first needed a model to describe the freezing of the fluid. For this purpose, we created a finite element model for the progression of the freezing front in the fluid and verified this based on literature data.

On this basis we developed a system model to investigate the freezing of the entire system (tank, pump and hoses) under various boundary conditions. We were then able to determine the required number and position of elastic elements in the overall system, such that the increase in volume due to the phase change did not result in an excessively high pressure in any part of the system.

With these two models we were able to resolve the conflict between the elasticity of the pump for freezing robustness and the rigidity of the pump for accuracy.

The key success factor was our focus on developing a model-based design concept with agreed modeling depth. The work was not without detours, setbacks and difficulties. The decisive factor, however, was that we as a team were able to improve our understanding of the system by gradually refining our modeling approach. This made it possible to develop a concept and a design within a few months, which was later produced in millions.

The engineering leader in the above example established the following list of expectations. The list has proven to be beneficial in practice over and over again:

1. All designs are modeled to an agreed depth.
   This means: no design without a model that describes the relevant variables. The agreed depth is based on the optimization target, the budget available for modeling and optimization, and the expected response time. For example, a mechanical element can be modeled with an analytical calculation and material data from the literature or using FEA[1] models with statistically verified material data from tests.
2. Accuracy and limits of the models are documented.
   This means that the accuracy and limits of the models used are known and suitable for the application. Without knowledge of accuracy, a model is not reliable.
3. Models are kept up-to-date and consistent throughout the life cycle of the product.
   This requirement means that decisions in product maintenance are also made based on models. If new developments and product lifecycle management are located in different organizations, the implementation may be challenging.
4. Excellent organizations work consistently on advancing modeling depth.
   Modelling is not a one-off business, but a constant effort. New design processes, new tools, new materials mean that modeling skills must be constantly further developed. In many areas of work, teams achieve the ability to design more precisely by improving the modeling depth. A more precise design is almost always associated with a competitive advantage. Excellent organizations therefore recognize modeling as a way to stay ahead of the competition.

## 10.3   Model-Based Engineering in Practice

As trainers, we observe that many organizations struggle to implement model-based development consistently. Procedures, which have been critical to success in the past, such as the frequent design optimization of prototypes in late stages of development, are difficult to overcome. If you want to implement the expectations described above, you can consider the following steps:

1. Analyze the situation: Which development steps are already consistently model-based or could be model-based? Where are critical gaps in modeling? Which competences are missing? This analysis is tricky since employees who work predominantly test based tend to underestimate the maturity and availability of models.
2. Plan and execute the implementation of competencies, resources, models and procedures systematically. This includes especially a competence buildup for engineering leaders. They should be able to commission models and to review results.

---

[1]Finite Element Analysis.

3. Reduce test capacities in parallel to growing capabilities in model-based engineering. Focal points of the experimental activities are shifting from a "product engineering by experiment" to "increasing the efficiency in the verification". Without a shift of resources from testing to modeling, there is no sufficient efficiency gain.
4. Create a roadmap for further development of modeling approaches. The "important" progress in modeling often competes with other "urgent" tasks and therefore needs special attention from leaders.

This change needs to be accompanied by intensive communication with the managers and employees. Scrapping a test facility, ending an established work pattern, possibly even transferring design responsibility to other teams, is not easy for anyone. In practice, the entire change can be expected to take approximately one product development cycle. In addition to properly planning for this change, it is important to prevent relapses into old ways of working.

▶ **Practical Tips**
- Always ask in a design review for the essential influencing parameters. Ask which effects should be considered and which can be neglected.
- Ask for suitable models and discuss them.
- Ensure reusability and reuse of models, e.g. through easily understandable model documentation and accessibility.
- Create a roadmap for building modeling capability and set targets for your employees at every step.
- Reduce the resources for experimental work as much as possible.
- Be attentive and challenge believes ("I think that …") that favor test based engineering work.

---

**The Most Important in Brief**

Model-based engineering helps to reduce development risks and accelerates product development.

Consistent application means that there is no design without modeling. In order to achieve efficiency, this needs to be combined with a reduction in test capacity.

Changing to a culture of consistent model-based engineering happens in small steps based on trust in the models.

Keeping models up to date over product lifetime supports product updates, e.g. in case of quality issues.

# Keeping the Product Promise

**11**

> ▶ In this chapter we show how functional and non-functional product charac-
> teristics are planned and realized in early phases of product engineering,
> even with high product complexity. We present "Design-for-X" as a universal
> approach that can be applied to all quality attributes.

Our products have to meet many expectations. Product developers often focus on specific product functions, e.g. a cordless screwdriver should operate for a long time and
re-charge quickly. Implicit or unspoken expectations are often missed for example, customers also expect a cordless screwdriver to:

- fulfill its function over the expected service life
- not hurt anyone, for example by flying parts
- be easy and convenient to use.

In addition to customer expectations, the manufacturer has expectations that:

- the manufacturing processes will be efficient.
- the product will strengthen the brand and motivate the purchase of further products—
  not only in the short term, but also in the long term.
- comply with laws, e.g. recyclability.

Other stakeholders have additional expectations in non-functional areas such as storage,
transport, packaging, labelling, etc.

New technologies and market trends constantly generate new opportunities.
Connectivity is a good example. Connected cordless screwdrivers enable an improved
management of tools on a construction site, when they provide online information on

© Springer-Verlag GmbH Germany, part of Springer Nature 2020                    67
M. Jantzer et al., *The Art of Engineering Leadership*,
https://doi.org/10.1007/978-3-662-60384-0_11

state of charge, and location. The control center can thus optimize the use of tools and provide sufficiently charged devices at the place of use.

Since the exact position can also help thieves to find their gain, this data must be protected. If the use is associated with the storage of personal data, they have to be protected as well. This drives new requirements for data security. This little example shows us how the prioritization of requirements changes over time. Until a few years ago, information security was not an issue for most products. Today, internet-of-things solutions have to manage who is allowed to access the product and who is not.

Quality models, e.g. ISO/IEC 25010 for Software Product Quality [1] give a structure and an overview to typical expectations. They classify requirements according to criteria such as functionality, reliability, safety, etc. (Fig. 11.1).

Each of these quality attributes is a typical starting point for a so-called "separation of concern", a design principle which breaks down the complex design task for a system into manageable "Design-for-X" sections: design-for-function, -reliability, -security etc.

## 11.1  The Design-for-X Approach

For each Design-for-X, a guideline can be created, and responsibility can be assigned to the team. With the Design-for-X approach, the overall design task can be structured: Thus, the large number of expectations (requirements) is broken down into multiple views, which are processed individually and afterwards integrated in the complete product. Step-by-step all the product promises are kept. Each Design-for-X runs through the entire V-cycle, i.e. it corresponds to the process model agreed in engineering.

The principles in Design-for-X are:

1. **Design so that the quality attribute X is met as specified.**
   The quantified and agreed specifications are the result of requirements engineering, which has all stakeholders in mind. Laws define the minimum requirements. User requirements are added. Laws do not always supply exactly the specifications needed for the design. Then ask for the intention of the law and from there derive the requirement.
2. **If 1. is not feasible, design to an acceptable quality level.**
   Legal requirements are not negotiable. For other requirements, however, an acceptable quality level may be agreed among the relevant stakeholders based on an opportunity and risk/threat assessment. Transparency about the premises for such negotiated quality levels is a prerequisite.
3. **If neither of 1. or 2. is possible, then do not design at all.**
4. **Observe the users in the field.**
   Statistical data from user observations are important for validating the assumptions made during the development phase. This provides valuable insights for current and future product generations.

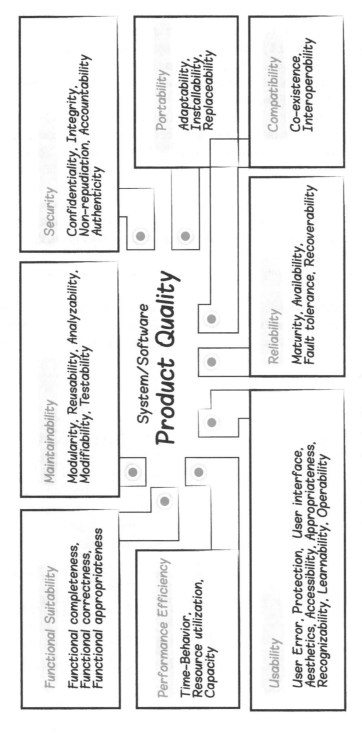

**Fig. 11.1** Quality models summarize functional and non-functional properties

**Design-for-Security** Using the cordless screwdriver as an example, we want to show how Design-for-Security can be implemented. Stakeholders, for example the operator of the cordless screwdrivers on a construction site, are sources for security requirements, such as protection against access to the position data of the individual devices.

The task in requirements engineering is to find specific threat scenarios for which designers can develop suitable protection solutions.

Solutions are often feasible on several system levels: Is it sufficient to protect the server on which the data is processed and stored, or does each cordless screwdriver itself have to be protected? In the case of Design-for-Security, the decision about measures to be implemented is the result of weighing up the resulting opportunities and threats (Fig. 11.2).

Product security is a Design-for-X discipline characterized by a constant emergence of new requirements based on news threats, e.g. hacker attacks. A connected cordless screwdriver system must be constantly protected against new attacks from the outside.

The basic design must enable these continuous adaptations. Suitable roles and processes have to be defined to support such adaptations; even after product delivery.

**Design-for-Reliability** Technical systems are often mechatronic systems in which software and hardware components interact. "Design-for-Reliability" aims to ensure that both perform their functions harmoniously and as expected over the service life of the product.

**Fig. 11.2** In Design-for-Security, threats are weighed according to potential risks and measures are addressed accordingly

**Software dependability**

Software dependability means the correct execution of a function over time. Since programmed code does not change during the life of the product, the focus is on designing a robust software architecture, error-free programming and cascaded test and integration processes (including unit tests, continuous build and continuous integration). The strategy for software dependability rests on four pillars:

- Avoidance of certain error classes in advance, e.g. by dispensing with non-transparent code, code re-use, and compliance with design rules.
- Early detection of errors, e.g. through peer reviews, automated tests and a formal check of the source code.
- Increased fault tolerance in the system, e.g. entering a safe state in the event of a fault.
- Fault prediction and classification. A classification can be used to decide whether the errors can be prevented, corrected, or tolerated.

**Hardware reliability**

Hardware properties change over the service life, e.g. because materials age, wear or are stressed. In order to guarantee the product's reliable function over the long term, we have to take into account the changes in properties during the service life.

A cordless screwdriver, for example, is subjected to a number of stresses on the construction site, e.g. forces, humidity, UV light, and temperature fluctuations. As materials expand with increasing temperature, dimensions change creating mechanical stresses. Temperatures change during the day, over seasons, and according to the operational load. As a result, the components experience cyclic mechanical stresses. This may lead to material fatigue and as a result, to a fatigue fracture and ultimately to a failure.

To prevent this, the loads (in this example it is the temperature change and number of cycles over the service life) must be known as well as the mechanism of fatigue that causes the damage. By using a model that maps the damage mechanism, a probability of failure can be determined. Statistics are important because parts, manufacturing processes, and material properties are distributed within tolerance limits. Achieving the promised reliability target depends on having carefully verified the accuracy of the reliability model.

If damage mechanisms are determined in advance of the actual product development, reliability can be designed into the product. In the early phases of development, there are greater degrees of freedom to divide internal loads in such a way that damage mechanisms are not even stimulated (Fig. 11.3). With our cordless screwdriver, for example, we could place a vulnerable design element at a position with a small temperature change. This would not activate fatigue in the first place.

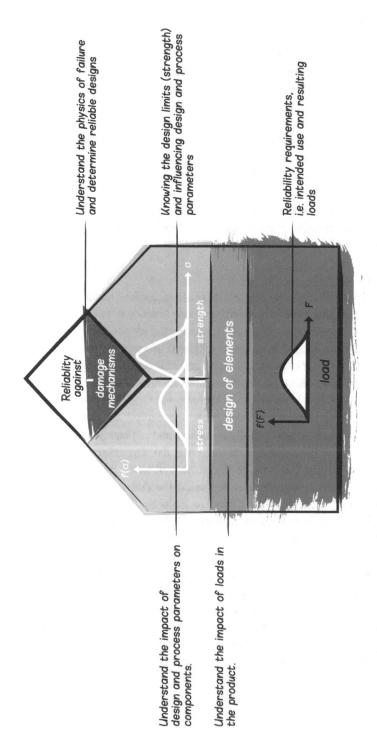

**Fig. 11.3** The tasks in reliability design represented as a house

The alternative approach of proving reliability by "testing" the finished product is usually more expensive and more time consuming. Typically, it takes a considerable time until product samples can be produced and tested. Any failures indicating that the necessary quality has not been achieved will require additional product samples. Higher reliability targets, i.e. lower permitted failure rates (e.g. for product safety), require even more parts for testing to enable statistical verification.

## 11.2   Leading Design for X

With "Design-for-Security" and "Design-for-Reliability" we have shown approaches for the implementation of two quality attributes as examples. The same applies to "Design-for-Manufacturing and Assembly", "Design-for-Product Safety", "Design-for-Cost", "Design-for-Test", etc. The leadership tasks are similar:

- Prioritize quality attributes: Quality attributes have dependencies and may be contradictory. A clear hierarchy of quality attributes gives the development teams orientation regarding how to solve design conflicts and where to spend effort.
- Set a design target such as a maximum failure probability for Design-for-Reliability. Usually stakeholders set targets. Law demands state-of-the-art failure probabilities for product safety. If there are no legal or customer specifications, economic considerations can be incorporated into the goals. In our opinion, it is favorable that senior executives set targets for quality attributes, as they significantly influence the perception of the company in the market and the development costs.
- Develop models for quality forecasting including verification strategies (see Chap. 10). Develop competence in the application of statistical methods. Performing the tasks requires experienced product developers who master every level of the V-model in their field of expertise. They should be able to collaborate with other domains to work out and decide on contradictory requirements.
- Clarify roles and power for decision-making for the respective experts. If done properly the experts will have a significant decision power. A good balance in decision making requires that the engineering leaders have sufficient understanding of the topic so that they are able to commission, to review and decide. Engineering leaders who lack this competence will not be able to utilize the full potential of their experts.
- Set up an appropriate review process and consistently adhere to it.

▶   **Practical Tips**
- Ensure that non-functional product properties are recorded as early as possible and are defined by the leaders.
- Build up your competence and your team's competence on the required Design-for-X topics to a sufficient level. Establish peer-reviews as standard.
- Review procedures as a central measure for achieving the quality goal.

**The Most Important in Brief**

The "Design-for-X" approach ensures that non-functional product properties are taken into account in the development process right from the start.

Leaders have the task to define and prioritize quality attributes of non-functional properties and to decide on contradictory requirements.

## Reference

1. ISO/IEC 25010:2011(E): Systems and software engineering – Systems and software Quality Requirements and Evaluation (SQuaRE) – System and software quality models. Geneva (2011)

# Leading Experts

**12**

> With the growing complexity in product development, the variety of experts
> required to reach or exceed the state of the art is increasing. On some topics
> this means that engineering leaders may even lack decision-making com-
> petence. As a result they have to relinquish a part of their leadership role to
> experts. We present approaches for effective involvement and strengthening
> of experts.

Leading experts is a special task, because experts often have a different set of priorities than leaders. Experts lead due to their knowledge without direct disciplinary responsibility. Appropriate involvement of experts in the decision-making process increases
competitiveness.

In general, with increasing complexity, more and more people with specialized
knowledge are needed who can effectively bring in their knowledge at the right time.
The cooperation models usually extend beyond the boundaries of one's organizational
unit and often even the company.

Let us take a completely different look at daily project life:

**Example**

"Appearance of the Expert"—a scene from the play "The Project"

The project manager steps into the conference room in complete disillusionment,
where ten people look expectantly in his direction. He pulls the expert by the collar
behind him with his last bit of strength. The expert fights back with his hands and
feet and calls out loud and outraged: "I'm not even here, I have an appointment with
the supplier, I urgently need to prepare for my conference lecture, and other customer
projects are also up against the wall!"

"Now that you are here, please stay just half an hour", the project manager begs
with folded hands and sounding somehow weepy and angry at the same time.

© Springer-Verlag GmbH Germany, part of Springer Nature 2020
M. Jantzer et al., *The Art of Engineering Leadership*,
https://doi.org/10.1007/978-3-662-60384-0_12

"All right, but only a quarter of an hour," the expert finally gives in as he straightens up and smoothes his jacket.

The others in the room—let's call them "the project team"—take a step back and nod with satisfaction. A murmur goes through the team. Noticeable sighs can be heard.

Now the project leader opens the discussion. A technical discussion develops between him and the team. The expert at first stands a little apart.

Then the project manager turns to him: "So, Mr. Expert, what do you think?" After an awkward pause, the expert starts his verdict: "The supplier uses the controlled-micro-injection-precision-supplement-process, which is state-of-the-art, by the way. This affects all tolerances. You can forget your previous assumptions. You have to trigger a combined-balanced-fluid-reaction-simulation, and then... then you'll see what's really happening."

It is deathly quiet. The ten people and the project manager take a step back. Some lower their heads. One begins to work on an imaginary spot on her blouse with embarrassment. Others look at each other questioningly. It feels like an eternity passes, and the restlessness grows. Finally, the project manager turns to a whiteboard in the background.

"Well, let's do it like this," he says and writes something down.

Meanwhile, the expert has already approached the exit at a running pace and calls back over his shoulder "So guys, now I really have to go, I hope it helped". The project manager asks pitifully: "You'll come back, won't you?", but the expert is already out of the earshot.

Admittedly, the scene is exaggerated and shows a dysfunctional expertise, but perhaps you have already experienced something similar. What can leaders do to prevent similar situations?

Becoming an expert often takes several years. It is not sufficient to attend a few specialist seminars, but rather it requires developing capabilities that combine deep theoretical knowledge with broad practical application experience (Fig. 12.1). However, even that is not enough. The expert must become recognized in his community as the go-to person for difficult problems.

Application experience is mainly passed on orally. Good experts are therefore often well connected with their peers. Exchange among other experts gives them new impulses, helps them to stay at the forefront in their field, and challenges them professionally. They motivate each other. Conferences or informal meetings are good opportunities to network (Fig. 12.2). In an efficiency-driven organization, where no direct value contribution arises from expert exchange, this freedom must be constantly encouraged and defended.

What makes a good expert? Expertise, exceptional problem-solving ability, and the aptitude for meticulous detailed work are required. Developing complex products together with experts from other domains, in other organizational units, and across continents, requires strong social and communication skills and the ability to change perspectives.

**Fig. 12.1** Competence is the combination of theoretical and practical knowledge

Experts must be able to present facts in an understandable way: verbally, in writing and by means of visualization. Only when complex interdependencies are modeled, i.e. so simplified that everyone else in the project can understand the reasoning of a decision, can a broader audience learn them. This creates the desired sustained learning, and multiplier effect. This means that the next time, the team can make a decision alone or at least develop proposals.

Knowledge from various fields must be combined in order to find and realize opportunities for innovative and sustainable solutions. Experts act as translators and mediators between the disciplines. Sometimes they have to explain the basics repeatedly, often for years and decades, because new employees and leaders join the organization. People who tend to be impatient, use esoteric terminology, or act like a know-it-all are completely out of place. Instead, patience and good will are favorable characteristics, ideally complemented by openness and intercultural competence (compare [2]).

In our experience, experts are motivated by the security of a long-term commitment from the organization, by their recognition as decision makers in their field, and by continuous opportunities for personal development. Leaders can strengthen this motivation

**Fig. 12.2** Knowledge is mainly exchanged orally. Networking supports the updating of expert knowledge

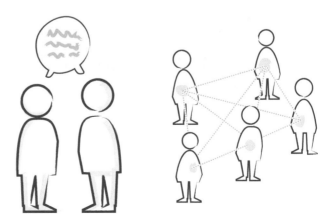

by highlighting experts' contributions to the achievement of objectives and consciously granting time and opportunities for networking [1] (Fig. 12.2). To boost the value contribution we recommend combining the expert's role with a design responsibility. For example, an acoustics expert has the responsibility to define and implement a design methodology that realizes the noise requirements of an appliance at minimal cost.

In a complex world in which markets, business models and technologies are changing ever faster, we need experts with special knowledge. Often we need an entire team of them. The model of "top down" leadership is not useful in this environment, instead we share leadership with the experts [2], i.e. responsibility and decision-making authority. Leaders may temporarily feel a loss of power and control, but in return they get commitment and sustainable decisions.

▶ **Practical Tips**
- Build up your expert organization systematically. Make sure that leadership skills are strengthened in addition to technical skills.
- Pay attention to your experts and provide opportunities to share new insights among colleagues.
- Create a common understanding of roles.
- Make sure that your experts are first and foremost technical leaders and only secondarily technical consultants.

**The Most Important in Brief**

Experts make complicated issues discussable.

Clarification of roles and responsibilities between line function and experts is an important building block for effectiveness and efficiency.

Experts perform leadership tasks. They make decisions and are responsible for them.

# References

1. Medinilla, A.: Agile Management, p. 75. Springer, Heidelberg (2012)
2. Kühl, S.: Laterales Führen. Springer, Heidelberg (2017)

# Risk Management

# 13

▶ In this chapter we present the objectives of risk management[1] and a maturity model to strengthen the focus on opportunities. Since risk management is a regular activity during the development of any product, we discuss process models which support early risk taking, utilization of opportunities, and proper reduction of threats. We also present practical tips for using FMEA and dealing with field problems.

In many areas it takes years until the success of development projects can be evaluated. Until success is achieved. There are many dangers lurking along the way.

- Will there be any unexpected technical difficulties along the way?
- Will the market accept the innovation at all?
- Will other companies reach the market first with a better solution?
- Are the economic assumptions still correct?

Risk management means consistently dealing with these questions right from the start.

If controllable threats are not systematically reduced and controllable opportunities are not systematically increased, the probability of success is reduced for the entire project. The consequence might be budget and schedule overruns, hidden mistakes, or even failure of the entire project.

---

[1] A risk may be an opportunity or threat.

© Springer-Verlag GmbH Germany, part of Springer Nature 2020
M. Jantzer et al., *The Art of Engineering Leadership*,
https://doi.org/10.1007/978-3-662-60384-0_13

**Example**

An example of such an approach is the construction of the Elbphilharmonie in Hamburg. An innovative construction project was awarded on the basis of unfinished plans. The successful bidder pointed out the massive risks associated with the unfinished planning. The client made special requests during the planning, but did not do any risk management.

From the first cost estimate until awarding the contract, project costs already rose from € 75 million to € 241 million [1]. During the construction phase, risks developed into serious problems, for example concerning the design of the roof and the fassade. In the end, this innovative project cost € 789 million and its completion was delayed from 2010 to 2017 [2]. Despite the cost overruns, substantial technical debt (incomplete work packages) remained. After just one season, substantial repair was needed.

Good risk control is a key to achieving challenging targets on innovation, timing or cost. Although risk management is nothing new, we see various degrees of maturity (see Table 13.1) in practice. The highest maturity an organization can achieve is a consistent opportunity orientation. To achieve any higher maturity level, the lower levels must be fully developed.

## 13.1    Prerequisites for Good Risk Management

Teams and leaders need the same understanding of the relationship between risk and innovation, and a common understanding about how to deal with risks during the project. Since development projects without innovation are a waste, development and risk are inextricably linked. In the early stages of product development, developers take high risks to take advantage of as many opportunities as possible. These risks are systematically reduced over the course of the project, as shown in Fig. 13.1. In doing so, the team pays attention to the risk preference of their stakeholders. They do not take any risks that stakeholders would consider unacceptable.

Another prerequisite is a culture of openness and risk transparency. Only if we discuss and evaluate risks openly can we achieve a good balance between market opportunities and technical risks. We would all like to develop a product that would become the innovation, cost, and quality leader in the market. The market risks for such a product would be virtually zero. Unfortunately, so are the chances of realizing this. Openness and risk transparency enable good risk-informed decisions. The deliberate concealment of risks in order to steer a decision in a certain direction is a serious breach of trust. This sometimes means that we have to bring an euphoric team back to reality. It is sad to lose some of their positive energy, but in the long term, it is the right decision.

**Table 13.1** Maturity model for risk management (based on E.M. Hall [3])

| Maturity: Phase: | Problem orientation | Solution orientation | Prevention orientation | Quantitative risk management | Opportunity orientation |
|---|---|---|---|---|---|
| Goal | We are tired of fighting fires | We want to know what can go wrong | We do not want to have regrets for not having acted in time | We want to know our probability for success | We want to realize the opportunities that arise |
| Discovery | We are too busy solving current problems to deal with potential problems | We are aware of the risks | We systematically investigate risks and understand their causes | We regularly evaluate prospects of success based on quantitatively assessed risks | We regularly and systematically identify the opportunities that allow us to achieve the optimum result |
| Processing | If problems arise, we will respond to them | We use fallback solutions if risks materialize | We reduce existing threats as planned and avoid problems | Current status and progress in risk work are transparent and allow the right measures to be taken in good time | We plan the realization of opportunities and follow through |
| Measurements | The pursuit of threats is pointless. We solve problems | We track project-threatening risks | We regularly monitor and evaluate our risks | We use risk triggers to detect plan deviations at an early stage | We document unused opportunities and lost profits |
| Improvement | In the next project we will invest more in frontloading | We learn from problems and ensure that we avoid them in the future | We reflect on misjudgments and promptly improve our approach | We create transparency and actively manage the project risks in order to take high risks in the early stages | We use the creativity of the entire team to exceed our goals |

The aim of risk management is to mitigate threats in a targeted manner and to exploit opportunities that arise

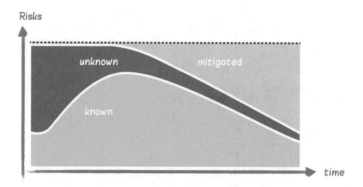

**Fig. 13.1**  Risk over time for a development project

Companies need a decision-making culture based on opportunities and threats. For each opportunity (probability of occurrence times potential gain) and each threat (probability of occurrence times potential loss) there is a point in time at which we must actively decide how to deal with it. If we delay the decision, then this often leads to an automatic loss of opportunity or an even higher threat. We are no longer managing risk, instead we are gambling. In our opinion, it is important to create a corporate culture that calls this game of chance by its name and does not celebrate the lucky players as heroes. Only then risk management can establish itself within the teams.

Finally, you need sufficient methodical and professional competencies as well as a suitable process to systematically discover, evaluate, and process risks.

## 13.2   Procedure in Risk Management

Often when we observe projects, the answer to our question on the state of risk management is "we do an FMEA". If we continue and ask about the status of the FMEA, we receive the answer: "We haven't started that yet. It's not due until the next milestone". This means that the project is missing many possibilities to identify and close risks at an early stage. As the project progresses, decisions and investments are being made. Reversing decisions becomes time consuming and expensive.

One reason why teams may not start systematic risk work early is that the start of risk management does not receive sufficient priority. In our experience, a potential first step is to create a structured overview of the system knowledge. This overview enables the setting of priorities for the development work. The functions of a selected hierarchy level in the system (e.g. level of submodules or components) are analyzed and categorized according to the following criteria:

C   The design for the implementation of a function is already verified and the inputs and outputs of the function are within the verified value range. All risks should be known and be at an acceptable level.

B    The design for the implementation of a function is already verified, but the inputs and outputs are partly outside the verified value range. These functions are classic candidates for future field failures, if they are not given enough attention. New problems can occur outside the proven work area. It is therefore important to examine possible risks at an early stage.

A    The work area is not completely known. There are knowledge gaps in these system elements. A functional failure cannot be excluded due to the lack of knowledge. Typically, most teams will focus on these functions on their own initiative, as they are aware that there could be great risks lurking here.

Risk management priorities can be derived directly from this map of system understanding. However, knowledge gaps do not necessarily lead to high risks, if there are proven backup solution or other remedies. Therefore, it is preferable to prioritize development according to the resulting risk and not directly to knowledge gaps.

The knowledge map is constantly updated over the course of development and expanded to include lower system levels. Knowledge gaps are identified, and priorities are set for the development team regarding any resulting risks. This is done according to the procedure model shown in Fig. 13.2.

Successful risk management always starts with a stakeholder agreement about the procedure. Rules and roles are clarified and the budget required for risk management is defined. The risk categories, evaluation principles, and permissible risk levels all must be

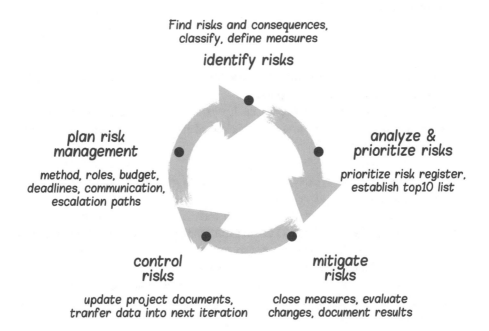

**Fig. 13.2** Incremental and iterative steps in risk management (following [4])

established and agreed. Some organizations try to replace this step with guidelines. Such an approach is not beneficial, as risk preference and stakeholder support remain unclear without direct interaction. Since stakeholder risk preferences can be volatile, it is important to maintain this dialogue on a regular basis.

In the second step, threats and opportunities are identified and a risk list is created. In addition to the risk category and the assigned responsible person, this list contains all the information currently available on this risk and its treatment. Since the biggest project risks are typically non-technical [5], the FMEA, for example, is only one source for the risk list. Discussing the risk list regularly is a core task for the stakeholders of a project.

When the risk list is available, threats and opportunities must be prioritized qualitatively for severity and urgency. The important ones are then quantified. It is very important to be cautious, since probabilities of occurrence and effects are very difficult to estimate accurately. Daniel Kahnemann [6] describes numerous examples in "Thinking fast and slow". The challenges include overvaluation of incomplete information, anchor heuristics, where people unconsciously judge their estimates by an anchor value, and unconsciously neglecting underlying distribution functions.

Based on the prioritized list a work plan is created. The aim of the work plan is to achieve the right level of risk at each project stage. This requires actively managing opportunities and threats using the following strategies:

- Change the project scope in order to mitigate threats or take advantage of opportunities.
- Reduce/increase the probability of occurrence or the impact, e.g. by building up knowledge.
- Transfer risk, e.g. by taking out insurance policies or by shifting risk to customers or suppliers by means of contracts, but also sharing opportunities through intensive cooperation with customers or suppliers.
- Passively accept risks with low impact, i.e. by waiting to see if there is a problem.
- Actively accept risks, i.e. by preparing backup plans that are triggered once a risk exceeds the acceptance limit.

In practice, we see that some strategies for dealing with risks are regularly misused and thus fail to achieve their objectives.

Changing the project content can reduce threats or exploit opportunities, but this requires a precise analysis of the risks arising from the changed content. New technical solutions in the design stage often look overly attractive at first glance, with the associated risks only becoming apparent after careful analysis.

In the case of technical risks, a transfer to suppliers is often attempted. However, the transfer of technical risks is only effective if the supplier has better opportunities or competencies to reduce the risks.

Passive acceptance as a strategy is regularly used for risks with a low probability of occurrence and severe effects. These individual risks are often low on the prioritized risk list and therefore are not evaluated quantitatively. If there is a sufficient number of these, the project is exposed to an unintentionally high risk, which remains unrecognized. Therefore, it makes sense to choose the passive acceptance strategy only for risks with a low impact.

The reduction of risk through backup plans and backup solutions regularly fails because the solutions are not adequately planned and developed. The rule of thumb is that backup plans and solutions must be proven and risk-free, otherwise they are not suitable.

The fifth and final step is to monitor the risks and any triggers that indicate the probability of occurrence has increased or decreased. When analyzing major quality issues, it is repeatedly noticed that no attention was paid to obvious triggers. When a projects encounters one failure in a test series, you often hear statements declaring "This was an isolated case". Any failure is of course a trigger, which means that the probability of a risk to materialize has increased. Dismissing a failure as an isolated case is a form of self-deception, which is one of the barriers to effective risk management.

## 13.3   Removing Impediments

Our experience has shown that the introduction and implementation of systematic risk management in many development organizations is difficult. There may be many reasons for this, e.g. inadequate training of risk management in engineering studies, or the perception by some employees and managers that risk management is a waste.

Leaders can demonstrate that it is useful for everyone if risks are identified early and, if necessary, closed.

The greatest endorsement for risk management is leadership that visibly treats it as a core task. Assigning responsibility for each threat and each opportunity, regularly discussing the risks, and actively managing risks show the importance to the team.

With suitable reviews, the risk tolerance within the organization can be shaped and balanced.

Another difficulty lies in the unproductive application of standard methods. As good and important as the FMEA is, the wrong emphasis can lead to bulky and marginally useful FMEAs. Here are a few examples of problematic approaches:

- Striving for completeness, FMEA moderators take note of every conceivable risk. The treatment of risks without practical relevance becomes tedious and leads to superficial discussions. This increases the probability of overlooking relevant risks, or relevant details, like exceeding functional limits.
- The FMEA is started too late, i.e. after all major design decisions have been made. Even if relevant risks are discovered, passive acceptance often remains the only

possible path. Thus, the work is perceived as a documentation exercise with very little value.

* The parties involved are technically or methodologically inexperienced, overlook relevant risks, and are unable to assess either the probability of occurrence or the impact in a meaningful way.

To be successful with FMEA, it is a good idea to start in the concept phase with experienced employees. In a project with a low degree of innovation, the existing FMEA of the predecessor product should be used as a starting point. If there is no predecessor product or no FMEA, the rough concept risks are evaluated first and then these are processed and refined during the project. Both procedures lead to a timely focus on relevant risks.

---

**Example**

In our consulting we have seen many good examples of how systematic quality work can be integrated organically into a project process. The project managers have adapted the FMEA tool exactly for their purposes in handling risk.

At a very early stage in the project, they mapped their system or product concepts into the FMEA and checked that they met the main requirements (functions and properties). Fields for which the team did not understand the cause-effect-relationships or had already identified tangible risks, the priorities were set to "dig deeper".

If the concept already contained verified components from previous systems, a check was made as to whether the verification was still valid under the new requirements, This included checking if external loads (e.g. temperature) did not exceed verified operation ranges.

The risks were prioritized using the risk priority figures. In the next step, the project team worked closely on these priorities. They identified cause-effect relationships, further clarified requirements, considered fallbacks, and adjusted the project planning.

Thus, over the course of the project, a step-by-step evaluation and processing was carried out along the focal points from rough to detail.

This approach provided the project team and stakeholders (including customers) with a high level of transparency about the current status of the project. It supported entrepreneurial risk management from the very beginning. The project proceeded efficiently and met its targets.

The FMEA sessions were interesting and useful for the developers. The content was relevant and gave a realistic picture of the current state of the project. The developers were able to act, decide, and bring their stakeholders on board where necessary. "I didn't think FMEA could be so much fun," commented one of the developers, adding, "we were in the driver seat".

When dealing with threats and opportunities we find very different types of stakeholders: from the experienced to the inexperienced, from the risk averse to the gamblers, and from the intuitive to the rationalist.

Daniel Kahnemann [6] has described the so-called priming effect, where the whole team follows the first opinion expressed. The so-called halo effect is where some people who are seldom questioned dominate everything through their personal effect. Often risk junkies have exactly this charisma, and spread optimism where caution is necessary.

Leaders can mitigate such effects by asking all decision-makers for individual assessments including their rationale.

▶ **Practical Tips**
- Discuss the threats and opportunities regularly with the project team and get actively involved in the risk assessment. Carry out cost reduction workshops already during development in order not to miss any opportunities.
- Ensure that experienced employees carry out risk management and that evaluations from previous products are taken into account. Be sure to focus on relevant risks.
- Ensure that all participants submit their assessments when critical issues arise. After all assessments are available, discussion follows and a joint decision is made.
- Understand your own risk preference and match it to the needs of your business.

**The Most Important in Brief**

Good risk management enables teams to take high technical risks in early phases and thus gain competitive advantages.

Risk management must be put in place right from the start. The evaluation should be carried out according to consistent standards. This can be achieved through regular discussions and requires a systematic approach.

A transparent handling of opportunities and threats leads to a corporate culture that enables innovation and avoids gambling.

# References

1. Buschhüter, O.T., Hamann, J.: Bürgerschaft der freien und Hansestadt Hamburg: Bericht des Parlamentarischen Untersuchungsausschusses „Elbphilharmonie", Drucksache 20/11500, Hamburg, 2014, p. 46. https://www.buergerschaft-hh.de/parldok. Accessed 1 Oct. 2018, 11:45
2. Kämpermann, M., Hotes, K.: Elbphilharmonie Hamburg. HamburgMusik gGmbH, Hamburg. https://www.elbphilharmonie.de/media/filer_public/54/57/5457895a-24e5-41dd-9ba2-852a5df-31bad/broschuere_elbphilharmonie_hamburg.pdf (2014). Accessed 1 Oct. 2018, 13:00

3. Hall, E.M.: Managing Risk: Methods for Software Systems Development, p. 60. Addison-Wesley, Boston (1997)
4. Project Management Institute: Practice Standard for Project Risk Management, Project Management Institute, Newton Square, Pennsylvania, pp. 9, 17 (2009)
5. Boehm, B., Lane, J.A., Kookmanojwong, S., Turner, R.: The Incremental Commitment Spiral Model, pp. 239–241. Addison-Wesley, Boston (2014)
6. Kahnemann, D.: Thinking Fast and Slow. Penguin, London (2011)

# Design Reviews

# 14

▶ In this chapter, we present the purpose of design reviews and what leaders can do to lead effectively by content through design reviews. To this end, we introduce the method DRBFM (Design Review Based on Failure Mode).

Leadership is always about compiling information in the form of premises, requirements, intermediate results, etc. in order to make decisions. The knowledge is in the minds of developers, in CAD (Computer aided Design) or simulation models, and programming codes. Each discipline uses its own type of information processing: model sketches, drawings, calculation formulas, flowcharts, etc. for visualization. With the help of these visualization tools the results can be jointly evaluated in reviews and improved until they are sufficiently mature.

By "Design Reviews" we mean all kinds of technical discussions that help to solve a development task together in a team. The objects of reviews are design, process and manufacturing drafts of different levels of detail and maturity. Design Reviews accompany the entire product development process. By contrast, project reviews are more about project progress, resources and budgets.

Design reviews jointly assess whether the current design will meet the requirements. If this is not the case, further measures are necessary. Developers frequently conduct such reviews when they coordinate with customers, other developers, suppliers, manufacturers, etc. (Fig. 14.1).

It can be useful to establish a standard for design reviews, especially for complex technical systems and if multiple parties are involved. This ensures the quality of the results and creates efficiency, because all participants can find their way into the reviews and zoom-in to the "essential points" quickly. The higher the quality requirements and the shorter the development times, the more important are efficient reviews.

© Springer-Verlag GmbH Germany, part of Springer Nature 2020
M. Jantzer et al., *The Art of Engineering Leadership*,
https://doi.org/10.1007/978-3-662-60384-0_14

**Fig. 14.1**  Design reviews support product improvement

Design review standards take into account the consistency of the technical system, that is, the relationship of the review object to other levels and components of the overall system.

## 14.1    The DRBFM Method

In addition to FMEA (Failure Mode and Effects Analysis), we have gained extensive experience with the "Design Review based on Failure Mode", DRBFM. DRBFM was developed at Toyota around 1997 [1] after an internal analysis showed the limits of FMEA work. A highly formalized effort was observed, but technical discussions were not conducted in sufficient detail at critical points. In addition, Toyota attributed many quality problems specifically to changes in products and processes.

Tatsuhiko Yoshimura, a reliability engineer at Toyota, had already introduced FMEA. Now he was looking for a simple method to eliminate mistakes in product changes by using the existing knowledge. At the same time he wanted to stimulate creativity and visualize the object of the review. Toyota did not see either of these adequately in the FMEA.

DRBFM starts with changes to the design, functions or requirements. These changes are analyzed in detail. The resulting potential failure modes are identified and examined. Then developers evaluate the resulting potential failures ("concerns") in a joint design review and eliminate them.

While an FMEA covers an entire product or system (and often has to cover it for legal reasons), in DRBFM developers are allowed to focus on specific areas of a design. The method deliberately dispenses with too many formalities. Instead, a design section is examined and improved in a well-prepared expert discussion, i.e. the review.

Toyota also realized that the introduction of a new method alone would not solve the problem. That is why Toyota embedded DRBFM as a concept in its quality culture of

Mizenboushi. Mizenboushi is a concept of preventive quality assurance and aims to prevent product defects before they occur. This avoids high additional costs and an extension of the development process beyond the start of production.

The pillars of Mizenboushi are the "GD³", which means "Good Design", "Good Dissection" and "Good Discussion":

- "Good Design" means an existing, proven design, which is well thought-out. Any changes to it are questioned. It also means that developers should always strive for a robust design and do everything possible to prevent errors from occurring in the engineering process.
- "Good Dissection" examines and visualizes the design model of the developer. The guiding questions are: How did the developer proceed with the design? Why has it been changed and which error chains could result? All premises and assumptions of a design are made transparent and can be critically and positively questioned.
- In "Good Discussion", potential risks are openly addressed. They are jointly evaluated and potential measures are discussed.

Mizenboushi and the GD³ are integrated into Kaizen, the Japanese philosophy of life and work, which requires continuous and infinite striving for improvement ("Kai" = improvement, "Zen" = for the better). Kaizen, Mizenboushi and GD³ are fundamental values and mindsets in DRBFM (Fig. 14.2).

In the first step of DRBFM, a scope is defined. The developer can concentrate on individual components, junctions, software or hardware components, system or subsystem functions. The scope results, for example, from changes to an existing design. Even with new developments, parts of existing designs are usually taken over.

In a second step, for the defined scope of the review, all functions performed by components and their interfaces are documented. If possible, assign component functions

**Fig. 14.2** DRBFM and Mizenboushi are embedded in the basic cultural mindset of Kaizen

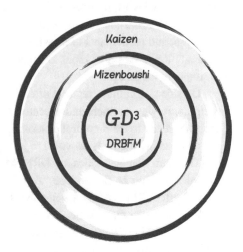

to system functions. Functions are carriers of requirements and define the scope for the reviewed design concept.

In the third step, the design is thoroughly examined for its behavior in all phases of the product life cycle such as assembly, testing, storage, transport, installation at the customer's site, operation and service. Ideally use visualized models for the behavior.

The aim is to detect possible weaknesses and eliminate the resulting risks or failures (step four). Root causes for potential failures are investigated. These are design or process parameters that cause a breach of functional limits. These parameters usually also indicate how the failure can be avoided in advance.

The last step is the definition of all necessary measures to avoid the potential failures. Toyota foresees the actual design review in this fifth step. Everything else is prepared by the responsible product engineer. At Toyota, the measures are at the core of the review.

---

**Example**

A few years after its introduction at Toyota, Bosch began to form an interest in this method and gradually introduced it to the company in order to improve design quality. Because Bosch has a different cultural background than Toyota, many things had to be translated and adapted to the Bosch culture. This took several years and may not even be completed to this day. One observation was that Bosch initially had great difficulty in focusing on small scopes in the reviews. In the beginning, the scope was often defined much too broadly. This was probably due to the familiar and legally required comprehensive approach of FMEA.

The review documents were more like wallpaper. They offered little overview and generated little added value to the FMEA. Increasing practical experience, growing management confidence, and the further development of the method to a "Modular Bosch DRBFM" have largely removed this obstacle.

Further barriers to using DRBFM were often the reluctance of developers to subject their own work to a "good dissection". In our opinion, this is part of the cultural differences: Kaizen is part of the daily life in Japan. It is natural to constantly improve a design, and participants of a review exercise their universally accepted kaizen "obligation". Reviewers give input to support the designer who welcomes the opportunity to improve his design.

Outside this culture, reviews often are searches for mistakes, gaps, weak points or culprits. Review partners and executives often exert additional pressure with a critical posture, ready to exert control or looking to attack. Then neither openness nor creativity arise, rather it comes above all to justifying and defending one's point of view.

The core characteristics of DRBFM—creativity and minimal formalism—were initially difficult at Bosch. The "Bosch modular DRBFM" significantly reduced the formal effort. This greatly accelerated its further spread, so that DRBFM is now used systematically in many areas. Through the consistent training of developers and leaders in DRBFM not only as a method, but also to improve the review culture, a mindset of "improve instead of prove" has been established.

## 14.2    Design Reviews as a Leadership Tool

Design reviews collect relevant information in order to discuss and improve the design. These form the basis for conscious decisions and make them a very effective leadership tool. Nevertheless this very efficient way of working is often neglected in daily business. It requires discipline, especially under time pressure when leaders rather look for action and fast decisions.

We have compiled a list of enablers and guiding questions, which we have concluded from our consulting practice to be useful for all types of reviews (Fig. 14.3).

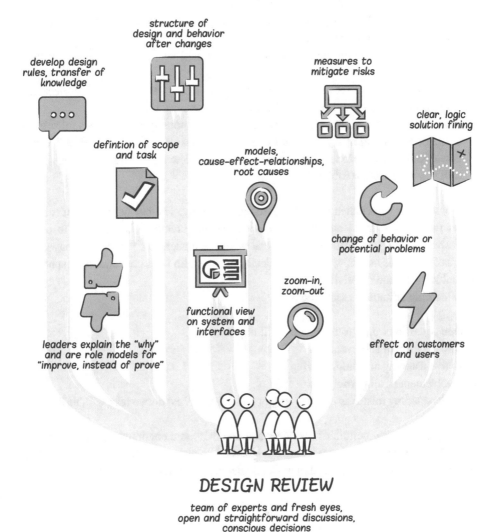

**Fig. 14.3**  Success factors for effective and efficient design reviews

- For each review, define a goal, the expected value contribution, and the scope of the analysis.
- 5–7 participants are ideal.
- Ensure that the review is staffed cross-functional with participants who bring new perspectives to improve the product.
- Clearly describe the subject: "How is it made (constructed, programmed)" and "What does it do, how it behaves (in the field, in operation)".
- Which functions should be fulfilled and how are they linked to the functional structure of the overall system?
- Model the functional relationships and behaviors.
- Clearly describe the change compared to an earlier design version. Is the design itself changing, the requirements, or both?
- What effects or potential problems are expected, and what does this mean for users and customers?
- Always link the content discussed to the development task (e.g. overall project) and the target of the review. This gives the review a clear story line.
- Even if the following looks old-fashioned: paper documents and sketches (also handwritten), everything at a glance on the wall or on boards are much better suited than PowerPoint presentations. This allows a good overview and invites an active discussion about the review subject.

Reviews should be oriented towards continuous learning. The behavior of leaders determines the creation of an atmosphere of openness and creativity, as opposed to one of defensiveness and justification. It is crucial what the leaders do, how they behave, how they ask questions and how they make decisions. The aim is to create an atmosphere of mutual learning and creativity.

A learning attitude is expressed by:

- Open questions, e.g. "What can we still improve?"
- Appreciation of achievements
- Active listening and summarizing
- Involvement of all participants
- Understanding mistakes as a learning opportunity: "What do we take with us?"

Developing a review culture of openness and creativity is a continuous process. The key is regularity. Repetition creates internalization and changes habits. The learning attitude described above can be practiced in any collegial conversation—as daily Kaizen.

▶ **Practical Tips**
- A review should be limited to 90 min.
- Set a scope and clear goals for reviews.
- As a leader in the review, clarify your role and your value contribution in advance.
- At the end, reflect for about 10 min whether the review met the expectations. Improve your reviews step-by-step.
- Establish a common review method in your field of activity to create efficiency.

**The Most Important in Brief**

Reviews collect different perspectives, competencies and depth of experience in order to improve the result. Design reviews are an important leadership tool for sharing knowledge and experience, developing targeted solutions and support active risk management. The leader's role is to establish an atmosphere of openness and creativity and to anchor a culture of striving for continuous improvement.

## Reference

1. Toyota: Beginners Guide to DRBFM (Nov. 2005)

# Decide

# 15

▶ In this chapter we introduce a model for decision making. On what basis do we decide? How can we make the situation around a decision discussable, and thus make more conscious decisions?

Besides organizing, decision-making is the main activity of leaders. By making decisions, we can reduce or actively increase the complexity of the development process. How does decision-making actually work?

## 15.1 Decision Timing

The latest possible decision time is determined by the product development project schedule. It is calculated backwards from the expected project completion. Once the product concept has been sufficiently validated, the required development effort is known and the resulting work packages are brought into a chronological sequence. The critical paths of development are now planned roughly from the project completion to the current status along the timeline. For elements of development that are critical to success, alternative solutions are developed that involve less risk for implementation. From the consideration of these critical paths and their alternatives, points in time are identified where decisions have to be taken.

These decision points, called "walls of decision" cannot be overrun without seriously jeopardizing successful project completion. That is not always crystal clear to leaders. Especially when decisions are difficult, they are often postponed—while the project team is tasked to create more clarity. However, if the decision time has been carefully derived, a decision will be made on this day, because there will be no standstill at the level of the engineer. The programmer, the designer or the project manager will continue to work based on their assumption of the most likely (postponed) decision. In fact, a decision is

© Springer-Verlag GmbH Germany, part of Springer Nature 2020
M. Jantzer et al., *The Art of Engineering Leadership*,
https://doi.org/10.1007/978-3-662-60384-0_15

made—one that has not been actively taken by leadership and may even contradict their idea.

When leaders want to live up to their role as decision-makers, they have to take decisions On Time. When taking far reaching decisions they are well advised to evaluate whether the "wall of decision" has really been reached. If not, they can still use the time to increase their certainty to decide through further investigations.

When leaders are able to decide earlier, they can increase efficiency, because the team does not have to make any further risk prevention for the current increment.

## 15.2   Certainty to Decide

Decisions are always made with uncertainty. If everything is clear, there is no need to make a decision.

**Example**

If you go to a restaurant and take a sample of all the food and drinks on the menu, then place your order, you have not decided, you have chosen. You have eliminated any risk of the unknown.

If you go to an exotic specialty restaurant, where you do not understand the menu nor the staff, you can still order a dish or a drink by pointing to an item on the menu. However, you have no idea what surprise the kitchen will serve. This process is called guessing.

If you understand the staff, you can have them explain all about the menu. You receive information that improves your decision-making ability. Nevertheless, you don't know exactly what you'll get, but you can weigh what you'd rather try. You're in a decision-making situation. You're weighing: What do you think will taste better? Which "adventure" do you want to take?

Thus, when making a decision, one operates in a range from guessing to choosing. In between lies the area of weighing (Fig. 15.1). In this area you know something, but not everything—an uncertainty remains. Most decisions should be made in the area of weighing. Why? The probability of being right is already good to very good. It limits the uncertainty and parallel work in the team. This way, leadership creates effectiveness and efficiency. In addition, leaders support the speed and quality of development.

The degree of uncertainty in decision making in which leaders feel comfortable is very individual. We have met very risk-minded leaders who feel comfortable and secure in the area of guessing. We have met hesitant leaders who only feel comfortable in the area of selection. Both are neither good nor bad. However, we have learned that leaders should know their personal willingness to accept risk. Depending on what is being decided, the aim is to achieve a level of decision-making certainty that is appropriate to the scope and situation.

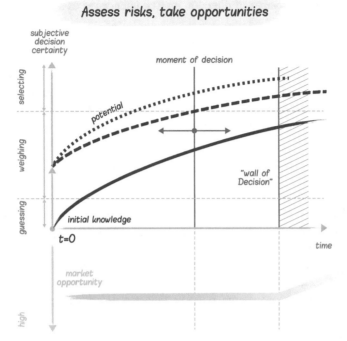

**Fig. 15.1** Decision-making

When deciding on a study to explore the next technology, taking a higher risk is likely to make sense—even to the point of guessing. Without risk there is no good prospect of a return. In this situation, decisions may be taken to challenge the team through a high degree of complexity and thus leading it to innovation.

When safety-relevant properties of the product are decided shortly before its market launch, low to no risks are desirable. Leaders should therefore strive for selection. This also applies to important and very large investment decisions. Guessing and gambling are close together. Who wants to gamble with the company? However, one should always bear in mind that a decision taken too late is an unwise decision. Typically, windows for market entry close over time. This is contrary to the quest for decision-making security. Therefore, deciding too late ultimately has the same consequences as deciding too early. This is visualized strikingly with the green line in (Fig. 15.1).

This shows that leaders have to be able to decide using the entire range from guessing to selection. Leaders should be aware of their own preference. They make decisions outside their comfort zone, if the situation demands it. If you are weighing it up, why not say it. If you are guessing, we strongly recommend that you say this as well. Because when making decisions with high uncertainties, subsequent changes of direction are likely. You must even plan to resubmit the decision and make corrections. In order to give the team a steady orientation, it is favorable to disclose the degree of uncertainty.

Otherwise, subsequent changes of direction may be understood as a lack of competence, fickleness or a lack of plan.

Especially in complicated or complex development projects, in which many people are involved and much is still unclear, discussion can increase the reliability of a decision. Perhaps a particular insight was assumed to be known and is longer applicable. Results may have already become obsolete or new ones are already available. The open dialogue creates more transparency and therefore a better basis for decisions. Resubmissions or the next decision milestones are thus planned in a target-oriented manner (also see [1, 2]).

---

**Example**

We once had a difficult product decision to make. One of our customers wanted a variant of a braking system that we had not planned. There were basically two options. The first was to use an existing valve family and rebuild everything around it—a complex and resource-intensive development task that would be relatively low-risk. The second option was a new valve variant, which would be technically very complex and was on the limit of what was basically possible. This required fewer resources. Moreover, this variant would make the product more sustainable because we could use it in other product variants.

Since the schedule was very tight, we, as a management team, tended towards the less risky option. Our decision reliability for the more attractive valve variant was too low. However, we deliberately asked the development team for its assessment of what they would recommend in consideration of all opportunities and threats. To our surprise, the team pleaded quite quickly and strongly for the new valve variant. Then they explained to us what made them so confident. We learned, what further thoughts and discussions the team already had, what additional findings were available, about opportunities with the valve and threats with the more conservative variant.

We then decided for the new valve variant. The development was anything but easy. In the end we were successful—not least because the team had proposed the decision and we all wanted to win. We finally finished the project "on time" and "on budget".

---

Openly discussing uncertainty in decision making has the additional advantage that newly gained insights are discussed thoroughly. This way new technical knowledge is spread and leaders are brought up to date. Above all, people typically don't talk about self-evident things, but what is self-evident to an individual is often not known to everyone else.

## 15.3   Utilize Your Time to Learn

Between today and the "wall of decision" is a time to learn and to build up one's subjective decision-making ability. This is a driving element of product development: understand the cause-effect relationships and make decisions for the best possible solutions based on them. In our coaching sessions we got to know experts and leaders who master the learning of cause-effect relationships with the available resources. They are focused on both increased understanding and product progress in product development. They are flexible in their learning strategy. When the upcoming decision is of great importance, they use "Design of Experiments" (or its virtual counterpart, "Design of Simulation") to make statistically verified statements. If time is more scarce, they conduct model experiments, analytical calculations, or combine several approaches. They are competent in their domains so that they can plan this reliably. They also use the time before the "wall of decision" for basic investigations of fallback solutions.

Successful leaders use company processes as a guideline that they adapt to actual boundary conditions. Others stick to the written process, even if they know that they cannot learn enough to make a good decision until the next milestone. They then require more time and more resources. From their point of view, this is quite understandable. They see the process as a premise for high quality whereas the former accept the constraints imposed by available resources and manage the risks in a flexible approach. Both approaches are based on positive values. As coaches, we have seen that the most promising way is to take conscious decisions on how you want to learn with your team. It also helps to talk about which learning has to be completed to take a decision, what is a given, what is flexible and tailorable?

Another way to learn quickly is to use existing knowledge. There is nothing faster than getting experienced colleagues on board. In larger companies, knowledge is very often found nearby, but also at universities or in technical books and reports. Even Internet platforms offer an astonishing amount of knowledge today although the reliability of the information source has to be checked.

There are very different approaches how humans acquire knowledge. We have met very well-read people, and communicative people who have large knowledge networks and those who try to learn everything by themselves. The flexible combination of all paths is certainly the most efficient way to acquire knowledge. And this combination also offers a good review culture according to our experience (see Chap. 14).

## 15.4   Three Factors in Decision Making: Head, Heart and Gut

If we ask leaders about their basis of decision-making, most of them answer: "facts and figures". After they have thought a little bit, they also name experience or gut feeling. Very few add: "I like the solution. It is elegant."

This leads to the empirical conclusion that people have three decision-making centers (as a metaphor):

- brain: the center for figures, data, facts
- guts: the center for learned patterns and non-explicable experience
- heart: the emotional center.

All three centers are involved in the decisions. Good decisions can only be made if all three decision-making centers are experienced as coherent: Are the cause-effect relationships understood enough, does the decision fit to successful solution patterns of the past, and does the solution touch us emotionally (Fig. 15.2).

If one of the decision centers is in conflict to the others, you should think again. For example: If the facts are correct and you fall in love with the solution, but have a bad gut feeling, then something is probably hidden and not yet properly understood. If you cannot fall in love with the solution, then there might be a better one.

---

**Example**

A few years ago, we had to evaluate whether it would make more sense to change the system architecture fundamentally for the next generation of one of our products, or whether the existing architecture should be adapted incrementally to meet future requirements.

We had carefully evaluated several scenarios and prepared the decision using QFD (Quality Function Deployment). We had compiled figures based on various scenarios. Almost all scenarios required a fundamental change in architecture. As a team, that seemed logical to us. In the presentation to our executives, however, we heard: "This is wrong. I can't tell you why, but that feels wrong."

Nobody wanted to leave it like this. We had invested a lot of time and the data was quite clear. Instead of arguing why we were right, we started an intensive discussion

**Fig. 15.2** Decision-making centers

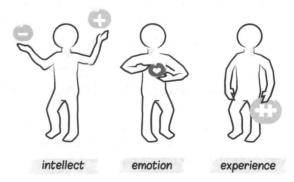

*All decision centers to be considered*

intellect            emotion            experience

about success patterns in the product, additional stakeholder requirements, and the elegance of solutions. The discussion was valuable and good. It took several iterations to lead us to a different more fitting architectural proposal. In the end, this proposal was simply more sustainable than the first fact-based concept.

In our coaching sessions, we consistently guide our leaders to conscious decision-making. We have often the impression that product developers believe that all their decisions are based on facts and systematic derivations that lead to decision matrices. The matrices then make the right decisions. On the surface this appears to be true, but when we honestly analyze how the "best result" is identified it is not strictly analytical. Experiences and preferences are integrated into weighting factors and individual evaluations. This means that even during the seemingly fact-based decision-making process, people synchronize their decision-making centers. Why not just disclose that? Is it not the discussion on decision-making that drives teams forward and allows them to make more conscious decisions?

When you take one more look at the three decision centers, you will see different time scales for the orthogonal decision centers heart, guts and brain. Emotional decisions are made spontaneously. Whether you like a solution and see it as elegant or not is usually immediately clear. Experience-based decisions can also be made very quickly. You have seen it before and you know it will work in the end, even if the concrete solution is not yet visible. Facts provide the most comprehensible basis for decision-making, but it often takes a long time before these facts are compiled and understood.

Here you can also see the advantage of knowledge and experience. They make product development effective, efficient and fast—while maintaining high development quality.

However, there is also a downside to experience and the recognition of patterns. They can hinder you when technologies change or you have to immerse yourself in other domains of knowledge. You will still be able to recognize the patterns you have known for a long time, but are they relevant now? We often see leaders moving from hardware development to software development. The leadership principles do not differ significantly, but the relevant technology-based patterns are different. These are initially difficult to perceive.

---

### Example

Have you ever been in the jungle before? Have you seen many animals in the first few days? Since our school days, we have known that the tropical jungle has the greatest biodiversity. You probably didn't see many of them the first few days, because you didn't know what the animals there look like. After several days and weeks, you slowly begin to recognize all the animals you didn't see in the first few days—the beauty and danger you hadn't noticed. You didn't see this initially because you didn't have pattern recognition yet. This is built piece by piece when you are attentive and curious.

## 15.5    Implementing Decisions

Do you also know this situation? Shortly before the end of a meeting, decisions must be made. Everybody has to leave for the next meeting. So, with the best of intentions you quickly summarize the essentials. Everybody nods and moves on. Just out the door, people start to have doubts. This tends to make last minute decisions unstable.

If leaders want to be effective, decisions must be implemented. Some decisions simply need to be enforced with no further discussion. This includes, for example, compliance issues.

There are decisions that must be translated into actions. If further derivations or even changes in behavior are necessary, the implementation of decisions is sometimes not so easy. In such cases, good leadership teams agree to regularly remind each other of the decisions made and their consequences. Such reminders are done in a friendly manner with the intention to serve each other. This mutual remembering and call to action supports success and personal growth.

Good leadership teams involve their subordinate units in the deployment process. Subordinate leaders are allowed to significantly influence the derivations of decisions. The leadership team makes an agreement on what is to be implemented and when. The participants remind each other of the given self-commitment. These teams grow together and commit to what they are able and willing to implement—and not what they dream of. In this way they sustainably develop themselves and their organizations. As a consequence they are perceived as effective.

A decision is only valuable if it is implemented and really changes something.

▶   **Practical Tips**
- Talk about your uncertainty to decide. If you have to decide based on a guess, then simply state it. Your employees may surprise you with further insights. They probably appreciate your openness.
- If your superiors cannot decide, but you can, talk about your decision-making ability and on what it is based.
- Carefully plan your "walls of decisions" and orient your project work toward decisions to be made.
- For important decisions, you should sleep on it one night.

> **The Most Important in Brief**
>
> Decisions are made latest at the "wall of decision" either explicitly by the leaders, or implicitly on the operative level (usually employees, suppliers or partners) whose processes continue.
>
> The time before the "wall of decision" should be filled with target-oriented learning.
>
> Good decisions are coherent in terms of facts, experience and emotion.
>
> Decisions with great uncertainty should be reviewed as soon as possible to ensure reliability.

# References

1. Storch, M.: Das Geheimnis kluger Entscheidungen. Piper, München (2015)
2. Kahneman, D.: Thinking Fast and Slow. Penguin, London (2011)

# Mastering Complexity

# 16

▶ The mastery of technical complexity is a core task in engineering. We show how complex problems can be transformed first into complicated and later into obvious tasks. We discuss the additional task of dealing with complexity of human behavior. In a kind of synthesis we then explore mastering the complexity associated with the development of modern mass production products, which is decisive for long-term market success.

Imagine the following problem. A customer wants to order a technical solution and you discuss requirements with him. During these discussions you find out that the customer does not yet have a clear idea about what he really wants. How do you proceed? Common sense demands: We must slowly approach the right solution.

Moving forward in small steps is a classic solution strategy for complex problems. We will clarify why this is a good strategy and which other strategies work as well. First let us clarify the nature of complexity.

In complex systems, the interactions between the system elements are not clear. We can therefore not predict the system response. Instead we can only observe it. Systems are complex if the system dynamics are high or the number of system elements is large or unknown. Complexity occurs in technical and social systems. Figure 16.1 shows typical complexity drivers.

Technical complexity is a subjective perception based on available knowledge about the system. The solution of a technical problem is complex for a team when they lack the necessary skills to find a solution analytically.

---

**Example**

The history of technology is full of examples of how inventors have solved complex problems. They developed methods and models that allowed them to master design processes and turned complex tasks into complicated design work. In 1896 Otto

© Springer-Verlag GmbH Germany, part of Springer Nature 2020
M. Jantzer et al., *The Art of Engineering Leadership*,
https://doi.org/10.1007/978-3-662-60384-0_16

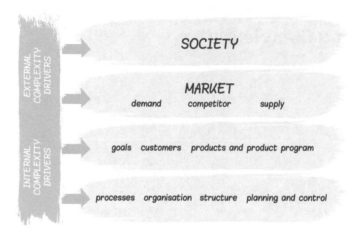

**Fig. 16.1** Internal and external complexity drivers in product engineering

Lilienthal paid with his life to achieve progress in the design of aircrafts. For the Wright brothers, the development of a glider based on the work of Lilienthal and others was already a partially analytically solvable problem. This allowed them to successfully perform their first powered flight in 1903 (see Fig. 16.2).

Shortly after in 1909, the technology was already developed to such an extent that a large number of aircrafts were designed. Step by step, an apparently unfulfillable dream became a complex problem and later a design task. Today students of engineering can learn it from textbooks.

To convert complex problems step by step into cause-effect relations that can be analyzed and modeled is the driver of progress. Engineers cannot effectively address new complex problems unless the current problems are easily solvable through design

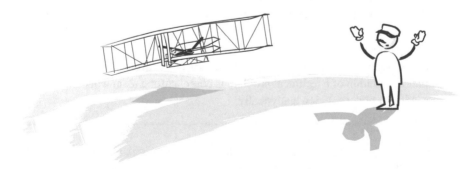

**Fig. 16.2** December 17th, 1903: Maiden flight of the "Flyer" developed by Wilbur and Orville Wright (Library of Congress)

guidelines and tools. Innovation means driving this process over and over again and mastering new technologies to create new products that better meet customer needs.

The purpose of leading your teams is not only to develop the next product but also to master the next technology, thus also securing the future of the company. If you focus your organization solely on the development of the next product without properly driving progress, you risk losing the power to innovate.

If teams regularly emphasize the "complex character" of their tasks, it is an indication that the innovation pace is slowing. In such a situation you can support the teams by reviewing the complexity of the task and their approach. The Cynefin model by David Snowden [1] described in Fig. 16.3 gives guidance in how to lead at different complexity levels.

The model divides tasks or problems into four categories: chaotic, complex, complicated and obvious. Chaotic situations are characterized by the fact that no correlation between cause and effect is recognized. The problem is incomprehensible and a solution is currently not possible. In such a situation, it is important to restore order. Once order is restored, a team may attempt to solve the problem. In product development, chaotic

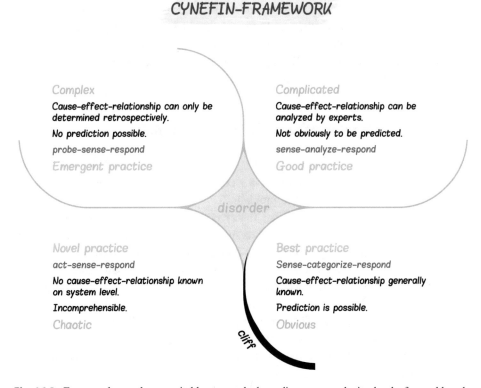

**CYNEFIN-FRAMEWORK**

**Complex**
*Cause-effect-relationship can only be determined retrospectively.*
*No prediction possible.*
*probe-sense-respond*
*Emergent practice*

**Complicated**
*Cause-effect-relationship can be analyzed by experts.*
*Not obviously to be predicted.*
*sense-analyze-respond*
*Good practice*

*disorder*

*Novel practice*
*act-sense-respond*
*No cause-effect-relationship known on system level.*
*Incomprehensible.*
*Chaotic*

*Best practice*
*Sense-categorize-respond*
*Cause-effect-relationship generally known.*
*Prediction is possible.*
*Obvious*

*Cliff*

**Fig. 16.3** Framework to select a suitable approach depending on complexity level of a problem by Snowden [1]

problems occur when things do not work as expected. For example, if many parts suddenly fail during the final inspection in mass production, a team cannot start the analysis of the problem right away. First, the team must ensure order and conditions under which the problem can be dealt with at all. The actions of the leaders should be characterized by "act—sense—respond" and aim to develop a novel practice. Complex problems are dealt with in small steps. You try something out, learn from it and then put what you have learned into practice ("probe—sense—respond"). New practices emerge stepwise. The try-outs become more and more productive. With a suitable domain expertise, complicated tasks can be solved analytically. Typically, this is also the most efficient way to solve them ("sense—analyze—respond"). Complicated tasks can be approached by several good practices. Obvious tasks are characterized by the fact that no analysis is necessary. The correct procedure is generally known ("sense—categorize—respond"). Leaders should strive for the best practice and implement it.

When leaders encounter new problems, they sometimes instinctively choose a known solution strategy ("disorder"). If the chosen approach does not fit to the problem, they run the risk of either becoming too slow (e.g. treating a complicated problem as complex) or ending in chaos (e.g. treating a complex problem as obvious).

In our work with teams, we start by classifying the upcoming tasks according to the Cynefin model. If a large part is located in the fields complex and complicated, we jointly consider the necessary steps to develop the problem step by step from complex to complicated and then to obvious. This is how engineering leaders get their organization's drive to innovate back on track. During the development of a new product, however, a certain amount of uncertainty will always remain.

## 16.1   Solving Problems that are Difficult to Define

"Wicked problems" [2] are characterized by the fact that defining of the task is itself already a problem. Product development in a competitive environment with finite development resources are typical "wicked problems".

Product development means solving future problems without the developers even knowing what the world will look like when the product is finished. In parallel competitors are working, or a start-up is turning the current market upside down with a groundbreaking invention. The external complexity drivers in Fig. 16.1 ensure that a great deal of uncertainty remains over the development period.

Experienced developers meet partly unknown system requirements with a robust, easily modifiable architecture. They already plan that they can change components and details with relatively little effort. On the other hand, an incremental-iterative approach helps them. They only work on the work products and specifications that are immediately necessary for the progress of the project. They can thus gradually reduce uncertainty and better identify threats and opportunities.

If requirements are defined too early and the architecture is designed too rigidly, there is a high risk that the product will later fail to meet the dynamic market requirements. Then extensive, expensive and possibly lengthy product changes become necessary. One example of this is the construction of the Berlin Airport.

---

**Example**

Construction of the Berlin Airport began in 2006 and was originally planned for 19 million passengers. Architecture and planning were based on extremely tight financial specifications from the states of Berlin and Brandenburg with planned costs of € 2 billion. Over the years, passenger forecasts have been massively raised, and the clients decided on various extensions. Since the architecture and technical planning did not foresee these typical airport extensions, there were massive cost overruns (>300%) and schedule overruns (>200%). Among other things, the entire cabling had to be replaced due to inadequate planning of the cable ducts.

---

## 16.2   The Human Factor

Development projects are not only about technology, but also about the people involved. After all, people make innovations possible through their creativity. They bring all kinds of professional and private goals to their projects. They make decisions based not only on facts, but also on emotions. They process information in completely different ways and deal individually with opportunities, threats and uncertainties [3, 4]. The leadership task is to maintain the necessary level of complexity that stimulates creativity, while reducing it to such an extent that a team can still work effectively on the common task.

Building trust is one way to reduce social complexity, because trust reduces uncertainty. Experienced leaders therefore spend a considerable amount of their time communicating with project teams and stakeholders. They create transparency, a common understanding of the objectives and achieve a focus on these targets.

Leaders can do even more. They can limit irrational behaviors in risk assessment and uncertainty avoidance by, for example, installing a transparent risk management process, by choosing robust solution concepts. Through short learning loops, the team can adapt to a fast changing environment. The team constantly improves and is thus in a better position to deal with uncertainties and to drive innovation.

Social complexity arises not only from people and from their behavior, but also from the way the people are organized. Unclear roles, distributed responsibilities, and unassigned responsibilities can block a development organization. Uncontrolled organizational complexity can be a trigger for structured organizational development (see Chap. 17).

## 16.3    Three Principles for Mastering Complexity

Anyone who has ever rummaged through hundreds of lines of poorly structured code during troubleshooting knows the feeling of powerlessness. In today's technical systems it is hardly possible or meaningful to consider all design elements simultaneously when finding a solution. Nevertheless, systems can be controlled by means of a few basic design principles.

One of them is "keep it simple". It is about finding the simplest solution to a problem. Whereby simply means: "the simplest manageable".

The second principle "divide et impera" has nothing to do with politics, but with the fact that a system can be controlled by dividing it into smaller units. Applied consistently, both principles lead to a modularization of the technical system. The modules are selected in such a way that they are as independent of each other as possible. This is also called "separation of concerns":

In order to develop modules concurrently, continuous integration into the system is required - the third principle. Continuous integration ensures that modules work across their interfaces. In the development of large software systems, this is achieved, for example, by integrating and testing the entire system every night. In mechanical systems, this is achieved by directly checking a newly created or modified component in the CAD system. We are here at the core of system structuring, which is described in detail in the chapter Architecture Design (see Chap. 9).

## 16.4    Limiting Complexity Costs

Modern technical products consist of a large number of individual parts. In addition, for various customer segments there are tailor-made variants. Simply try the configurator for buying a new car to get an idea of the multitude of potential individual vehicles. Cars can be ordered in so many different variations that identical combinations are rare. Mercedes, for example, has only produced two identical versions of the A-Class in a period of 5 years [5]. This poses a major challenge for the cost-effectiveness of mass produced products.

If diversity at product level would cost nothing and would not entail any effort in the life cycle, then any number of variants could be created. Every customer request could be met. However, variance has its price: marketing costs, logistics costs, set-up costs, purchasing costs and development costs increase, while economies of scale decrease.

So how do you keep variants manageable and optimize profit? Theoretically, this can be achieved by creating various scenarios with complete cost and market transparency and optimizing them. This allows you to determine profits related to your variant portfolio. Just like the variant portfolio, the component portfolio from which the products are built can be optimally structured. With such a modular strategy economies of scale can still be achieved even with great variance. However, because we do not know what the world will look like in the future, such a portfolio cannot be precisely optimized in advance.

We are therefore seeing constantly growing product portfolios. New niches are filled with new product variants. However, these new product variants do not necessarily win new customers who would otherwise have bought from the competition. It is often the case that similar variants weaken each other in sales success. This also means that important cost advantages are lost, because mass production products have strong price-volume dependencies. As a rule of thumb, doubling the number of identical units results in a 2–20% cost saving [6].

Variants also burden the development organization, since these new products not only have to be developed, but also supported throughout their life cycle. The complexity of the product portfolio often increases the complexity of processes and the organization too. This weakens the power to innovate. Portfolios must therefore be regularly adjusted in order to balance innovative strength and volume effects.

Portfolio management is a cross-functional task of marketing, sales, production, purchasing and development. Together they optimize the portfolio. By obeying a few rules and with some discipline, uncontrolled increase can be prevented and product complexity can be managed in a targeted manner:

1. Cost transparency avoids unnecessary conflicts. On the basis of willingness to pay, costs and quantities, the various domains in an organization find an agreement for portfolio extension conflicts.
2. Each system level takes responsibility for the complexity of its portfolio. This means that for example the battery development team of a car manufacturer suggests a portfolio of batteries and aligns a target portfolio with the management.
3. Modularization and standardization, e.g. of interfaces, minimize complexity costs through common part strategies. Variance at the module and component level must be controlled, preferably down to the last screw.
4. If modularization is also to work across several production lines or even production sites, it is essential to define a portfolio of permissible processes for each production process. Increases in part complexity due to incompatible production processes are inexpensive in the short term and expensive in the long term.
5. Complexity reduction is carried out in larger cuts. Only then is it cost-effective. Figure 16.4 shows classical strategies for this purpose. Every phase-in of a new platform is accompanied by the phase-out of the platform it replaces.
6. To make optimum use of economies of scale for cost advantages, project calculations must show to what extent new variants lead to the cannibalization of existing variants. The negative influence of reduced quantity effects must be minimized.

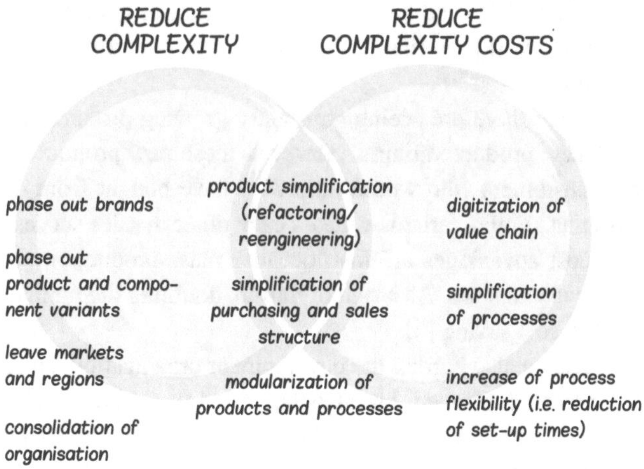

Fig. 16.4 Strategies for reducing complexity and reducing complexity costs according to Wilson and Perumal [7]

**REDUCE COMPLEXITY**

**REDUCE COMPLEXITY COSTS**

phase out brands

product simplification (refactoring/ reengineering)

digitization of value chain

phase out product and compo- nent variants

simplification of purchasing and sales structure

simplification of processes

leave markets and regions

consolidation of organisation

modularization of products and processes

increase of process flexibility (i.e. reduction of set-up times)

▶  **Practical Tips**

- Use the Cynefin model to categorize the tasks you and your teams are working on. Then derive the most suitable procedure.
- Clarify the complexity drivers in your task portfolio and make sure that each driver satisfies a customer need, otherwise eliminate it.
- Calculate the cost of complexity and compare it with the customer benefit.
- Pay attention to the controllability of your tasks and use the opportunities of modularization and hierarchical system design. Describe interfaces in appropriate detail.

---

**The Most Important in Brief**

Complex tasks require an incremental and iterative working mode, since these tasks cannot be solved analytically. Agile working modes are utilized.

The principles "keep it simple stupid", "separation of concerns" and "continuous integration" facilitate the handling of complex questions.

We counter social complexity by building trust, transparency and clear roles.

Modular systems and portfolio management on product level are tools for mastering the complexity of customizable mass production products.

---

# References

1. Snowden D.J., Boone M.E.: A leader's Framework for Decision Making. Harvard Business Review, Boston, (2007)
2. Rittel H.W, Webber M.M.: Planning problems are wicked problems, Polity, pp. 135–144 (1973)

3. Gigerenzer, G.: Gut Feelings. Viking, New York (2007)
4. Kahneman, D.: Thinking Fast and Slow. Penguin, London (2011)
5. Wehking K.H., Popp J.: Linked logistic concepts for future automobile manufacturing using innovative equipment, p. 539 in Bargende M, Reuss HC, Wiedemann J.: 16. Internationales Stuttgarter Symposium: Automobil- und Motorentechnik, Springer, Heidelberg 2016
6. Ehrlenspiel K., Wiewert A., Lindemann U., Mörtl M.: Kostengünstig Entwickeln und Konstruieren, 7th edn, p. 181. Springer, Heidelberg (2014)
7. Wilson, S.A., Perumal, A.: Waging War on Complexity Costs, p. 11. Mc Graw Hill, New York (2010)

# Shaping the Work Organization

**17**

▶ Within the scope of this chapter, we show which aspects should be considered in the context of organizational design and how organizations are successfully changed.

We explain how managers can use a model to analyze and shape the organizational culture.

---

**Example**

Mr. Brown is new in his position as department manager in the development of home appliances. At various moments in his first weeks, he already sees that the long-standing organization does not optimally support the current business. The adherence to delivery dates for start of production is extremely poor, and the entire organization is burdened with a massive workload. He reflects about which organizational changes are necessary.

Many managers are familiar with this problem, regardless of whether they are taking up a new position or taking a step back to review whether their organization is still effective and efficient.

An organization that may have been suitable in the past appears ill suited for future tasks or is already dysfunctional. Typical signs of dysfunction are:

- Ad hoc rescheduling of resources so frequently that increments cannot be completed.
- Frequent escalation of resource decisions to higher management levels.
- Interpersonal conflicts between project and line managers, or between project managers, because individual interests are enforced to the detriment of the overall system.

© Springer-Verlag GmbH Germany, part of Springer Nature 2020
M. Jantzer et al., *The Art of Engineering Leadership*,
https://doi.org/10.1007/978-3-662-60384-0_17

- No overall transparency of activities and workload.
- Resource buffers are created and disguised by organizational units.

If the organization shows these signs, it is necessary to reshape the work organization.

## 17.1   Organizational Development as a Leadership Task

Engineering organizations are the object of constant change. They serve to achieve a goal and lose their raison d'être when this goal has been achieved. Just as a team of mountaineers dissolves when they have climbed the summit and have returned to the valley.

An important premise for organizational development is therefore to abandon the idea of long-term stable organizations and to regard each organization as a temporary assignment of people to tasks (or vice versa). Under this premise, there is no "best" form of organization. Instead, there is a regular need to find suitable work organizations to achieve new goals.

Nevertheless, we keep seeing that managers shy away from adapting their development organizations to changes, even if the employees are already suffering from an inappropriate organization. We typically encounter the following situations:

- Leaders misjudge changes in the relevant environment, e.g. the competitive situation.
- Leaders prioritize short term value contribution much higher than sustaining further development of the organization, e.g. by delaying the further training of their employees in favor of project work.
- Managers fear that changes could jeopardize the performance of the organization.
- Leaders do not know how to tackle organizational design because they lack the expertise to do so.
- Leader's fear tackling the inevitable conflicts associated with reorganization.

Let us look at them in detail: Misjudgments of the relevant environment usually have a rather long lead-time between action and effect. It is therefore important to identify the early indicators for a necessary organizational change. Potential indicators among others are degree of innovation and time to market of competitors, long-term changes in market conditions that effect competitiveness.

Prioritizing the provision of services higher than organizational development often arises from the prejudice that organizational changes are complex, expensive and lengthy. However, the application of the Kaizen[1] principle of improvement in small steps makes it possible to carry out improvements without great effort and with very short payback times.

---

[1]Kaizen: Japanese principle of "infinite change for the better", selected, gradual perfecting, e.g. continuous improvement.

The fear of a slump in performance of an organization is surprisingly widespread. It is clear that an inappropriate organization is demotivating. Angel Medinilla strikingly describes [1] that unmotivated employees never develop excellent products. In addition, the risk of an engineering organization collapsing due to reorganization, e.g. due to incomplete implementation or non-addressed tasks, is low. The informal organization, i.e. the network of employees, is usually so strong that organizations are not thrown off track immediately after incomplete or bad organizational changes.

Very often, however, the reason for overdue organizational development is the lack of competence of the leader. Only a few engineering leaders acquire sufficient competence in organizational development in the course of their studies and further training. Many managers therefore have difficulties in planning and implementing organizational changes in a targeted manner.

## 17.2   Organizational Development as a System Design Task

Probably the easiest approach to organizational development is to consider it as a system design task.

Organizations create value from inputs in a transformation process and therefore share some aspects with a technical system. At the same time, a development organization is a social system[2] with unimposing quiet people who create enormous value and fulfil tasks of central importance, critical obstructionists ensure that an organization never becomes saturated and satisfied, and lively empathetic networkers who hold an entire international organization together.

As with any technical system, the leadership team needs a common system model with system boundaries, inputs and outputs, and system elements that are interrelated. The Burke & Litwin [2] model in Fig. 17.1 describes these system elements with a focus on their impact on an organization's performance.

It is important to know the influence of the system elements on the system output. This can serve as an orientation for prioritizing measures in organizational development. The system model of Pflaum and Weissenberger-Eibl [3] in Fig. 17.2, which is based on the models of Negele [4] and Wenzel [6], divides the product development system into an enabling system, a value generating system and a context system. In the enabling system the prerequisites for development performance are laid. Engineering work, e.g. product engineering, is carried out in the value-creating system. The context system summarizes the external boundary conditions for product engineering. This model is particularly suitable for the analysis and optimization of matrix organizations.

---

[2]After Collin Powell: "Organization doesn't really accomplish anything. Plans don't accomplish anything, either. Theories of management don't much matter. Endeavors succeed or fail because of the people involved."

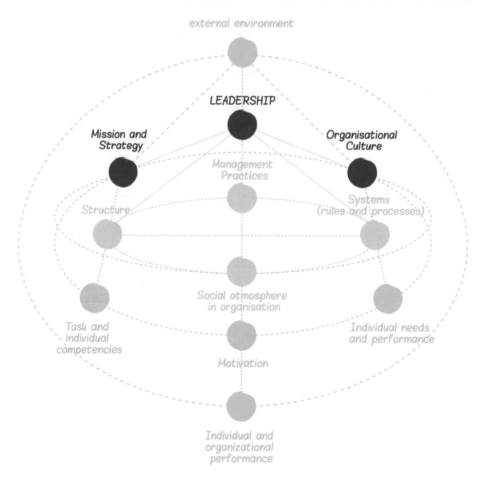

**Fig. 17.1**  Organizational model by Burke & Litwin [2]

Pflaum and Weissenberger-Eibl show which factors have a particularly high influence on the performance of the engineering organization and should therefore be the object of special attention in organizational development (see Fig. 17.2). The success factors on the left side of Fig. 17.2 can be measured straight away in the early phase of projects. This allows drawing conclusions early about the prospects of success of development projects.

The system models of Burke and Litwin and Pflaum emphasize different aspects. They represent different views helping to avoid misunderstandings and thus enable joint agreements.

We assume that organization design is always a joint work of the organization and not the lonely task of one boss. Our experience shows that it is beneficial to involve all those primarily affected by the resulting organization in the elaboration of the design itself, even if a certain inertia can sometimes be felt. Finding an agreement with the affected

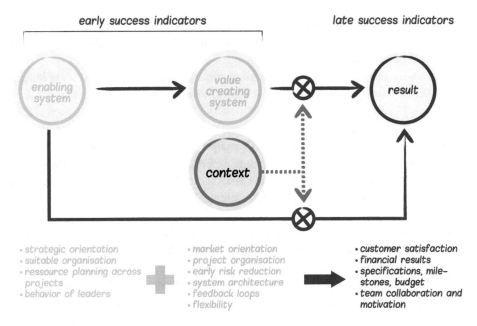

**Fig. 17.2** Simplified system model according Pflaum [3]

organization considerably reduces the risk of not achieving sustainable changes. In addition, it allows to integrate perspectives from different organizational levels. The joint system design with the team motivates and prevents a drift into a victim role. A wider number of participants also increases the chance of aligning the informal and formal organization in the end.

Let us return to our fictitious example and consider the first two steps of goal definition and modeling.

---

**Example**

Mr. Brown and his staff use a PESTEL[3] analysis as a simple tool for determining the situation. By this, changes in the relevant environment can be determined in a structured manner to evaluate their effects on the organization:

Among other things, they ascertain:

- The number of connected appliances is increasing. Consumers expect that future home appliances will make their lives easier for them through connected functions. This will lead to a massive increase in the share of electronics in home appliances.

---

[3]PESTEL: Political, Economic, Social, Technological, Environmental, legal—Collection of socio-economic search fields in order to evaluate systematically the relevant environmental changes for a company. Originally developed by Francis J. Aguilar [5].

- The global middle class has a growing interest in sustainability. This leads to increased requirements on environmental compatibility in production, use, and disposal of home appliances.
- Additive manufacturing processes allow new designs with high functional integration and low manufacturing costs at the same time.

In addition to a common view of the outside world, the group also needs a common view of its organization. In a SWOT[4] analysis, Mr. Brown and his team identify the strengths and weaknesses of the organization. As a next step they reflect on the company targets and the results of the analysis to derive targets for the department. Among other things, they set themselves targets[5] for

- The improvement of the adherence to promised delivery dates for new products
- The penetration of their products with connected functions
- The phasing out of certain substances in the products and the further development of the design methodology for the use of additive manufactured components
- Improving customer orientation, which was rated poor in the self-assessment of employees.

They develop a strategy to achieve these goals. They discuss it with their partners at the system boundaries and the leaders in the superordinate system. They then decide to reorganize their work organization and processes.

## 17.3    Requirements and Architecture Drivers

The first step in the system design of an organization is to determine the requirements, see Fig. 17.3.

In order to ensure the performance of a development organization, it makes sense to proceed with requirements engineering in a similar way as in a product engineering project. Stakeholders must be identified. The stakeholders include not just the recipients of services but also for example, superior management, human resources, the workers' council, employees, related functions, and so on. These stakeholders' requirements must be elicited, analyzed and evaluated.

On the other hand, the interests of the social system must be satisfied. One example of this is that employees strive for appreciation, autonomy and mastery in their work.

---

[4]SWOT: Strengths, Weaknesses, Opportunities, and Threats – simple tool for rough analysis of the organization's situation.

[5]SMART—Specific, Measurable, Accepted, Reasonable, Time-related. A quality model for target formulation.

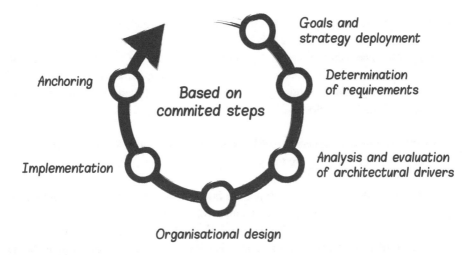

**Fig. 17.3** Tasks and phases in the system design of an organization based on Stanford [7]

This means that the value contribution of individuals within the organization must remain visible, that employees feel sufficiently empowered and have the opportunity to educate themselves regularly.

Once a first set of requirements is available, the next question is which of these requirements determine the work organization. Some of the requirements will have a major impact on the design of the organization. These architecture drivers require special attention. A typical architecture driver is the degree of complexity of the tasks. The higher the degree of uncertainty, the less the work packages can be planned in the long term and the easier a change of direction needs to be. To work with great uncertainty, very flexible work organizations are necessary. For this purpose, it is helpful when teams have only a few external interfaces and all competences are in the team.

In complicated fields of work, analysis capability and domain expertise are the keys to success. It therefore makes sense to bring together employees with the same domain expertise. They can learn from each other and build up more knowledge together.

Predominantly obvious tasks can be well planned. Since the tasks are obvious, the focus of the work organization is on efficient workflows.

Let us look again at our example:

---

**Example**

Mr. Brown and his team have identified the various sources of requirements. In addition to the neighboring departments from which e.g. services are purchased, these sources are the executives, the employees and of course the customers and partners of the department. The elicited requirements contain the following points:

- Executive management wants the organization to perform at least one project with a high degree of innovation and two medium-sized projects with a low degree of innovation in parallel.

- Employees expect an organization that provides a good work-life balance and ensures that the standard weekly hours are not regularly exceeded.
- They also expect from their management a clear strategic orientation in order to enable employees to make the right decisions deploying the strategy.
- Manufacturing as a neighboring function expects new product and manufacturing concepts to be co-developed.

At the end of this comprehensive analysis, Mr. Brown identified three decisive architectural drivers for the work organization:

- The new organization needs to support a fast knowledge built up in all things digital. New employees and experienced employees should learn from each other.
- Improving lead times and adherence to delivery dates is a crucial requirement. In the past, delivery pressure led to a massive overload of employees, which should be overcome. Since these requirements are contradictory, they must be considered together.
- Running parallel projects with different degree of uncertainty.

The number of architecture drivers for a work organization is typically higher than shown in the example. From our experience, it makes sense to focus initially on a handful of architecture drivers.

## 17.4    Architectural Design and Organizational Development

The next step is to design a new organization according to the architecture drivers. For the discussion of potential solutions the agreed views are used. As in system design, it is possible to create several organizational designs "top down", bring them to the same level of detail and then check them against the architecture drivers. It is important to check the robustness of the architecture against foreseeable changes in these requirements.

The comparison of different architectures against each other can never be objective. Despite all transparency and common evaluation criteria, it remains a question of defining an uncertain future on incomplete data. The tolerance of the participants towards uncertainty or concerns about their own role in the new organization mean that the assessments will always be subjective and biased by beliefs. These personal beliefs of the leaders and employees involved are legitimate. After all, the employees should be motivated to work in the future organization.

Let us take a look at this next step in our example:

---

**Example**

The architecture drivers are a real headache for Mr. Brown and his team.

- The necessary further development of know-how for digitization requires an organization in which employees can learn from each other. To this end, employees with the same domain expertise should work together.
- The parallel processing of several projects in a tight schedule requires a strong project organization with interdisciplinary and cross-functional working methods. The classic organizational architecture for these tasks is a matrix organization. New forms of organization for this are mixtures of agile teams and excellence clusters, which represent a professional home.
- The degree of uncertainty in the projects varies greatly. While innovative projects often require a change of direction, projects with a low degree of innovation have few uncertainties.

---

To run demanding innovative projects with extensive build-up of know-how in parallel with standard projects, the architecture must allow flexible work organizations. Mr. Brown and his team are designing different approaches for a new organization. They evaluate a complete agile organization with Scrum teams and different forms of matrix architectures including the current organization. Finally, Mr. Brown opts for an organizational model with permanently staffed project teams. They work lean or agile depending on the situation.

The teams composition is based among other things on communication density, analyzed through the Design Structure Matrix Method (DSM) [8]. The team structures and the assignment of tasks are optimized in such a way that communication-intensive tasks are handled within the teams. The necessary communication between the teams is minimized.

Teams can adapt their working methods to the tasks at hand. The Lean Principles are implemented in all teams. For tasks with low uncertainty, the teams work in a lean framework. When complicated or complex tasks (see Fig. 16.3) have to be mastered, the teams organize themselves in an agile framework.

To ensure know-how is built up, it is determined that the team members jointly review their results in the line organization and define special projects for the development and introduction of methods.

With these agreements, Mr. Brown and his team can now start to work out the details of the organization. First they set up a rough value stream for various development activities, see [9]. Inputs, outputs, tasks and their throughput times and dependencies become visible on process maps. The team encounters one or two problems. Some processes do not work well. These are collected and possible solutions are proposed. Using the previously determined value stream, the employees estimate the scope of the tasks per team, define the team sizes and start the assignment of the employees to teams and tasks. After some iterations, in which the team composition is optimized with regard to the characters and roles, a coherent picture emerges that Mr. Brown and his team agree upon.

As you can see from the example, organizational development is an iterative process.

## 17.5    Sustainable Implementation

There are many opportunities to fail in organizational development, especially in implementation and embedding. The main reason for this is that the leaders are not pulling in the same direction. We have observed the following in our work with engineering leaders:

- The introduction of the organizational change is not planned, its effectiveness is not measured, and progress is not reviewed.
- Senior executives are asking to coordinate the change with all possible parties in the company. During alignments, the draft is then watered down to the point where the new organization looks strikingly like the old.
- Formal approval of the change is followed by a reinterpretation of the content and a unilateral implementation aimed at safeguarding the interests of part of the organization.
- Organizational changes are fully explained and formally rolled out, but simply ignored by the organization. New processes are not applied because the old procedures, processes and tools are not stopped.

Probably you can think of a few more possibilities to fail at this point of organizational development. More interesting is the question why they fail and what can be done about it. Implementing change takes time. The more hierarchical levels an organization has and the more people have gained the experience that prior changes were only half-heartedly pursued, the more time implementation takes. Therefore, it is important to follow a detailed plan with achievable intermediate goals, for example with a PDCA[6] system. The model described in Fig. 17.4 is a simple analysis tool to determine which impediments have to be removed.

To ensure success, it is not only important that employees understand that the change makes sense and that they accept it emotionally. They must also have the necessary implementation skills and the power to implement the necessary change. The decisive factor is that they experience "successes". The more comprehensive and long-term an organizational development is, the more important it is that the teams draw new strength

---

[6]Plan, Do, Check, Act: Systematic approach to the implementation of targets.

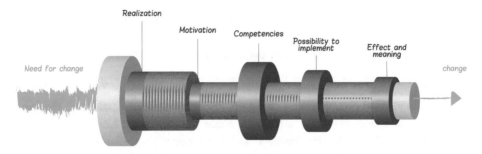

**Fig. 17.4**  The pipeline model [10] as an analysis tool on the status of the organizational change: The narrowest point determines how much change will happen

repeatedly from small and large successes. They need the support of the superordinate leaders and cross-functions. To ensure that the progress does not fade away or that the change as a whole is not called into question, an organization needs regular assurance that the path it has taken is the right one. Nothing is better suited for this than success and nothing is more dangerous than the idea that things no longer need to be discussed, since they have been decided.

The procedure in small steps is a simple means of responding to unexpected responses from the social system. In many cases, it is not easy to see whether the goals of organizational development are being achieved. In some areas success comes with a long leadtime. Whether a new car will be a great market success is known about 5 years after the start of development. Market acceptance is therefore not a useful indicator of the success of an organizational development measure. At this point, we come back to the aforementioned studies on success factors in product development (see Fig. 17.2).

In the model of Pflaum and Weissenberger-Eibl the following elements of the value enabler system are used as early success indicators:

- strategic orientation,
- appropriate organization,
- cross-project resource planning
- behavior of leaders

In the value creation system the success factors are:

- market orientation
- project organization,
- early risk reduction,
- system architecture,
- feedback loops
- flexibility

These indicators should be sampled frequently. We recommend a measurement and evaluation at least every six months.

Let us take a last look at Mr. Brown and his team:

---

**Example**

Once the planned organizational change is discussed with all employees, the teams draw up their own organizational plans. Since the implementation is taking place in parallel with the current business and synchronously in all teams, Mr. Brown draws up a plan for the organizational change. The individual steps are spread over 12 months, starting with the change in team composition in the first month. This is followed the introduction of joint technical reviews and all other changes to roles, rules and processes. Once a month the groups measure the success factors to ensure that they are on a good way. After a few month it becomes apparent that the changes are not working equally well in all teams and the motivation of the employees is decreasing. Therefore, Mr. Brown introduces regular reviews on the status of organizational development with all teams and their leaders. The plan for the organizational change seems not to fit. He notices that some employees in the new structure are still working according to the old procedures. In an interview, one of his specialists explains to him: "If I had known that you were serious about the change, I would have already started to implement the individual changes every month, but you must understand this was never important in the past".

Mr. Brown discusses this impediment with his employees using the pipeline model in Fig. 17.4 to get a clear picture where the organizational change is stuck.

Together they realize that the coexistence of new official rules and old unwritten rules leads to some employees being overloaded and subsequently demotivated. They then jointly define a few simple rules that allow these employees to reject old working methods and escalate these cases quickly. As a result, employee motivation is increasing and the other success factors are slowly increasing. For Mr. Brown this is the signal that he can start thinking about how to further improve the flow in his organization.

---

## 17.6  Developing Organizational Culture

Corporate culture is one foundation for successful product development. According to Hofstede [11], corporate culture describes a collective phenomenon that encompasses common values, norms, attitudes, basic assumptions, knowledge and traditions. Only a small part of it becomes visible in the form of actions, most of it remains hidden. Yet, these rules of the game and common patterns of action ensure smooth cooperation.

*"A few years ago, I moved to another company. This was a real culture shock"* is something you sometimes hear from friends and colleagues. *"I was totally overdressed in the first meeting with a jacket and tie"*, *"suddenly I had to address everyone*

*formally"*, *"in the new company I am not allowed to decide anything anymore"*.
However, after such a move, it usually takes only weeks or months, then the shock is
over and the new culture has been adopted. We usually adapt quickly, because constantly
offending and not being part of it makes you unhappy in the long run.

A corporate or organizational culture is the result of a process of development and
adaptation to the boundary conditions, often lasting for years. In order to establish common values, standards, rules and patterns, many alignments are necessary and many conflicts must be resolved. Thus, organizational culture also always reflects the history of an
organization, i.e. how it has adapted to the requirements and conditions of the past. In
doing so, it is also constantly in interaction with the cultures of its surroundings, e.g. the
culture of the country.

## 17.7   Shaping Culture as a Leadership Task

In product development, the boundary conditions can change very quickly. With globali-
zation, accelerating communication, and the emergence of new competitors, the markets
are changing rapidly. New products must be developed much faster. Additional pressure
to change is generated, for example, by a technological change such as transfer from
combustion to electric drive, or the change in the labor market due to demographic
effects. The prevailing culture quickly becomes a competitive disadvantage, if it does not
adapt.

The resulting leadership task is to develop a learning organization that can adapt
quickly. The art is to set the right pace and direction. On the one hand, high adaptability
is advantageous. On the other hand, changes that are too fast or too frequent contain the
risk that skills are not developed well enough or that communication channels cannot
develop stably.

A "conservative" culture scores with stability, security, and predictability, both for the
people in the system and those that are external. Competencies can be perfected in a stable culture. However, in the face of change, a stable culture can seem sluggish.

Leaders are the creators of organizational culture. The first step is to examine the prevailing culture in order to decide what should actually change. Values and standards are
difficult to grasp. However, leaders can easily observe what people in their environment
say and do. They can easily observe which beliefs are heard repeatedly, e.g. "we will
solve whatever problems arise", as opposed to "this will never work".

What people say and what they do is often not the same. Take the time to consciously
look at how people behave in your environment, e.g. in stressful situations, such as decisions or conflicts. When the pressure rises, people fall into their mostly unconscious
patterns. Here their own "bias" plays a role. You can then compare your observations
and decide at which point you want to start. It is helpful to take a systemic approach to

cultural design. This means keeping an eye on the entire social system and the system environment:

- Develop a common vision and mission that describes what the organization's objective and purpose are. These are linked to a common identity in which individual identities can be found.
- Involve those employees who already experience the need for change and are therefore motivated to implement it.
- Find common values and connect them with the common vision and mission.
- Observe the formal relationships (e.g. role relationships) and recognize the importance of informal relationships. Good relationships create stability, cohesion, and commitment.
- Consider the culture of the society in which a company is embedded. If the desired changes conflict with the broader external culture, the changes cannot be implemented at the desired speed.

## 17.8   A Model to Shape Cultural Change

The system characteristics shown in the previous section can be found in the "Levels of Learning and Change" (Fig. 17.5). They are based on the levels of learning from the social scientist G. Bateson [12] and were later developed into a communication model by R. Dilts [13]. This makes it one of many suitable models for describing and shaping organizational culture.

**Fig. 17.5** The logical levels [13] as an action model for cultural design

Figure 17.5 depicts a logical model and not a process model. Each level describes characteristics of culture and offers approaches to change. The levels are arranged hierarchically according to effectiveness in facilitating change. To change a level, the starting point is at least one level above it.

The model is a proven tool in coaching and change management, since it enables discussion about the "invisible" dimensions of culture, i.e. all levels above the "behavioral level". This opens up new options for leaders.

At the very top is the orientation towards a superordinate purpose. Here you will find visionary objectives with social relevance (protection of nature and environment, improvement of standards of living, health for many, etc.).

The identity corresponds to the self-image, e.g. "market leader", "innovation leader" or to a list of the most important beliefs: *"As market leader you have to be able to react quickly and cost-effectively"*.

Values describe what is important in a company or team, such as "openness", "sustainability" or "legality": *"Sustainability and legality secure the medium-term future of our company"*.

The level of roles and rules includes explicit and implicit rules and principles: *"We do not compromise on safety"*.

Beliefs express what employees and managers think about themselves, their colleagues, the team or the company. We discover them in sayings like *"This has never worked for us before"*.

The next level describes the knowledge, skills and competences and basic procedures, for example the way new products are developed and brought to market.

The level of behavior describes interaction between humans, i.e. with customers, colleagues, bosses or employees.

Environment describes the working environment where the interaction happens such as offices and laboratories, production facilities, working and communication equipment.

Successful change initiatives always have all levels in view. If they are initiated exclusively at the lower levels, e.g. at the capability level with "compulsory training" or through new processes or roles, but there is little or no communication and adaptation of purpose, identity or values, the changes can take too long before lasting effectiveness is achieved, and the initiative may lose momentum.

▶ **Practical Tips**

- Consult people outside your own organization to get an objective view of your current situation.
- Involve your employees in the design of the organizational change. They are able to find a coherent solution.
- Identify, set and track observable criteria for change.
- Take small steps and follow them consistently.
- Watch out for impediments to change, such as the following beliefs: *"That's how we do it, it works"*.
- Keep questioning yourself. Be careful not to become routine-blinded.

**The Most Important in Brief**

Organizations must constantly adapt to new tasks and constraints. The starting point is a systematic approach based on an analysis of the current situation. Drivers of change and other requirements are elicited and form the basis for the organizational design.

Good organizational designs emerge from the people within the organization itself. Leaders integrate their ideas and make them involved co-creators.

Sustainable implementation is based on good planning, monitoring, evaluation and regular communication.

The model of the logical levels helps to address all dimensions of cultural change and to involve the people by discussing and negotiating what is often left unsaid.

# References

1. Medinilla A.: Agile Management, p. 70. Springer, Heidelberg (2012)
2. Burke, W.W., Litwin, G.H.A.: Causal model of organisation performance and change. J. Manag. **18**(3):523–545 (1992)
3. Pflaum, B., Weissenberger-Eibl, M.: A Using network analysis to evaluate success factors for new product development, IEEE European Technology and Engineering Management Summit. München (2017)
4. Negele, H., Fricke, E., Igenbergs, E.: ZOPH- A systematic approach to the modeling of product development systems. INCOSE Int. Symp. **7**(1), 266–273 (1997)
5. Aguilar, F.J.: Scanning the Business Environment. Macmillan, New York (1967)
6. Wenzel, S., Igenbergs, E., Michl, T., Megerle, F.: Coupling Changes to Product-, Process-, and Agent-System Architectures EuSEC 2000, p. 129–134. Herbert Utz Verlag, München (2000)
7. Stanford, N.: Guide to Organization Design, p. 86. Profite Books Ltd, London (2007)
8. Yassine, A., Braha, D.: Complex concurrent engineering and the design structure matrix method. Concur. Eng. **11**(3):165–176 (2003)
9. Osterling, M.K.: Value Stream Mapping. Mc Graw Hill, USA (2014)
10. Stober, D.R., Grant, A.M. (Eds.): Evidence Based Coaching Handbook: Putting Best Practices to Work With Your Clients, p. 54. Wiley, Hoboken, NJ (2006)
11. Hofstede, G.: Culture and Consequences: Comparing Values, Behaviors, Institutions and Organizations across Nations. Thousand Oaks, New Delhi (2001)
12. Bateson, G.: Steps to an Ecology of Mind, pp. 279–308. University of Chicago Press, Chicago (2000)
13. Dilts, R.: Changing Beliefs with NLP. Meta Publications U.S., Santa Crus, CA (1990)

# Roles in Engineering

# 18

▶ In this chapter, we discuss the importance of role design for the best possible delivery of value contributions to customers. Building on this, we present a simple model about dimensions of leadership. We describe how to increase effectiveness by applying it.

## 18.1 Developing Roles

Honestly, how often do you actively work on understanding and shaping roles and the relationships between roles? When you go to the coffee kitchen and hear the conversations there, what are they about? For the most part, your colleagues and employees talk about relationships—about roles, role expectations and interaction between roles.

**Example**

Let's make a short excursion into another domain: acting. Roles are immediately understandable there. Movies depict clear roles played by individual actors. Each role fulfills a specific task in the movie. You know James Bond? What's his job? He has to protect the world from villains and their "dark organizations". His specific task is to identify and eliminate them—directly on sight. To achieve this, he has been given special authority, the "license to kill". This authority is linked to responsibility. He must defeat evil and protect the uninvolved from harm in the best possible way. He needs skills for that. For the role of James Bond, these skills are martial arts, charm, decisiveness and trauma resistance. In addition, he needs access to information—from his own sources, but also from those of his opponent.

© Springer-Verlag GmbH Germany, part of Springer Nature 2020
M. Jantzer et al., *The Art of Engineering Leadership*,
https://doi.org/10.1007/978-3-662-60384-0_18

Now, back to engineering. We do not want to make a movie. Rather, we want to develop products and services for our customers. We therefore start role definitions from the perspective of value contribution and the tasks derived from it.

Let us look at an example of a role in agile development: the Product Owner. The product owner plans the incremental design and delivery of value for the customer. The product owner divides the planned value contribution into smaller increments and formulates sub-goals, for example as user stories (see [1, 2]). He is therefore responsible for the sequence of value contributions in terms of content and timing. He has the authority to fill the backlog with increments, to set their priority, and to release the increments realized by the team. Only if he has these powers, he can take responsibility for the overall result. To do this professionally, he needs domain expertise, experience in product design, understanding of the customer needs, and knowledge of the company's boundary conditions, i.e. the internal stakeholder needs. (Fig. 18.1).

The product owner should have solid competences in agile requirements engineering, good communication skills and the ability to break down and commission engineering tasks. However, he does not have to be able to solve complicated differential equations. A team member can probably do that better.

## 18.2  Role Relationships

Roles have both an individual aspect and a relational aspect. In acting or in the movies, role relationships are clearly defined. They are an essential part of the story and are "enforced" by the director. In real life, enforcing relationships does not work, nor does it make sense. Very often we have seen in our leadership training courses that roles and role relationships are described formally. This is a good thing, if roles stay stable over time. However, it is not enough because roles can be interpreted. According to our experience, insisting on a static understanding of a role does not really help. We have often observed that a role owner is not able to fulfill the formal role. This can be due to the skills of the role owner, but very often is due to the social system in which the role owner is embedded. In both intercultural and cross-functional cooperation, interpretation of

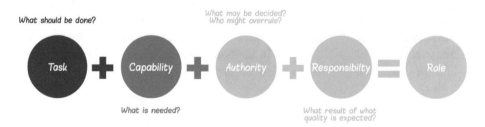

**Fig. 18.1** Description of roles

one's role and the expectations of others regarding the role often diverge. If the actual role relationships are different than intended or expected, this is perceived as dissonance.

It is therefore advantageous to design the role relationships with the respective role owners in order to generate the optimal value contribution for customers. To this end, those affected explain their own understanding of their role, express their expectations to colleagues, and negotiate explicit agreements (Fig. 18.2). Role clarification works best using concrete examples. According to our experience, specific terms are filled subjectively and culturally with different content. We see this above all in intercultural cooperation. Thus, different cultures interpret "taking responsibility" differently. If you do not become more specific, you leave too much room for misinterpretation.

In addition to the explicit role descriptions, there are unspoken role expectations ("I would expect a department head ..."). Talking to each other without reproach, listening openly, seeking cultural understanding, and "going to gemba" usually helps if we have the feeling that roles are not adequately assumed. Unclear role relationships lead to gaps in providing services or to conflicts and thus to loss of performance. From our experience, a purely formal assignment of roles to people is unfavorable because opportunities are not used and flexibility suffers.

In our view, roles should be assigned according to people's strengths. However, it must be ensured that the value contributions are generated.

The coordination of roles and mutual role expectations is particularly important at organizational interfaces. In functional organizations, these are the interfaces between functions like, Development—Purchasing, Development—Manufacturing, etc. In product-oriented organizations, these are the limits of the product lines. In our experience, organizational principles that are oriented towards different targets meet at these borders. Roles can no longer be easily transferred from A to B, and responsibilities are difficult to understand from the outside, because you look at other areas with your own view of the world.

Large purchasing organizations may be structured according to material groups that strive for the lowest integral purchasing costs of their material groups. The development project manager strives for the lowest cost of his product. The goals of purchasing and development can now be orthogonal to each other. For purchasing, it can be

**Fig. 18.2** Mutual alignment of roles

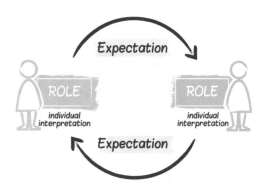

advantageous to transfer the business to a supplier who is not working to capacity, thus enabling a better overall contract for the company. However, the project manager may want to work with a well-known supplier who has superior technical competencies and has also made a good individual offer. Both parties optimize their view, and both views are valid.

If you want to achieve the best for your company, then consider such conflicting targets as an opportunity to win. Try to understand the interests of internal partners, uncover your interests and work out solutions for both interests. It is not about the developer or purchaser, but about the realization of customer and stakeholder needs.

You can remain true to your role description, while remaining open to win-win solutions with your partners. The best solutions are often those that in the beginning were neither yours nor your partner's.

## 18.3   Role as Leader

The career of an engineering leader can be based on various abilities. Domain expertise may be a starting point, organizing skills or the ability to make decisions. More and more we see that the ability to design systems and to develop individuals and teams is a trigger for advancing an engineering leadership career. We have developed a simple model to enable discussing the leadership role in development—for oneself and for others involved (see also [3]). We have set up a triangle with three generic roles "decision-maker", "expert" and "coach" (Fig. 18.3).

In the role of an expert, the leader supports his employees in carrying out tasks. He makes concrete contributions and suggestions. In this role, he shapes the system to reach or exceed the state of the art. It is fundamentally necessary that a leader has the

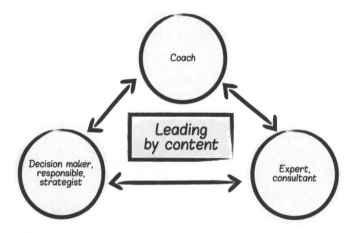

**Fig. 18.3**  Role model of technical leadership (by content)

"expertise to shape the development system". For example, the leader must be in a position to structure the work organization in his area of responsibility to maximize the value contribution. From our point of view, this design task cannot be delegated but needs a neutral perspective.

The role of the decision-maker is characterized by responsibility for a task. This can be a concrete development task, but also the development of (strategic) objectives and strategies. The decision-maker has the means and powers to accomplish this task. He makes the decisions and is responsible for the consequences.

The coach has the task of enabling teams and individuals to reach their full potential in their engineering work. It is important that the teams find their own way and take responsibility for it, even though the leaders stay accountable. This is a balancing act, which in our opinion, can only be achieved through a proven trusting relationship.

As leaders in product engineering, we can position ourselves at any point in the triangle of roles. There is not a right or wrong position, rather the situation determines a more suitable position. We have seen that the conscious reflection of the leader's roles creates more self-confidence and flexibility and thus more effectiveness in the end.

In Fig. 18.4 we modeled the dynamic interaction between leader and team as a kind of journey.

Imagine starting in a new area. On the first day your value contribution is zero, but also your effort. Because you are curious, you start asking questions. You are learning, but you are also holding up the team.

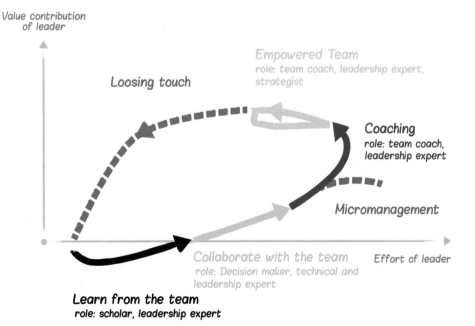

**Fig. 18.4** The spirit of technical leadership (by content)

Once you have learned enough, you will increasingly be able to take on the role of an expert and decision-maker. You will then typically still work very closely with the team. Some leaders stay there because the close interaction creates the feeling of immediate effectiveness. The step to micro-management is small.

If you want to be more effective, you need to empower the team to find their own way and make their own decisions. This makes teamwork faster (compare Fig. 6.3), and requires trust and the corresponding team skills. We do not think it means the team is running completely alone. You must stay in good contact or you will lose your effectiveness. It remains your task to point out the "big picture" again and again and to set the strategic direction.

Depending on the maturity of your teams and topics, you move flexibly on the curve. Team composition and boundary conditions change. It happens repeatedly that empowered working teams encounter such big impediments that you as a leader have to work intensively with the team again. As a leader, you typically have more room to maneuver and can remove impediments efficiently.

▶   **Practical Tips**
- Analyze whether a common understanding of roles exists in your area. Document agreements in writing.
- Check your role characteristics with regard to the maximum value contribution for your customer.
- Encourage the understanding of other roles and their value contribution by temporary job rotation or internal internships of your employees in neighboring areas.
- Use the three roles of the leader daily (e.g. in reviews) to create role clarity for yourself. Observe your effectiveness.

---

**The Most Important in Brief**

Roles and role relationships must be aligned to maximize the value contribution. Roles and role relationships should be explicitly agreed with all parties involved.

By consciously assuming the roles of "decision-maker", "expert" and "coach", the leader can increase his effectiveness according to the situation.

---

# References

1. Wintersteiger, A.: Scrum. Schnelleinstieg, 3rd edn. entwickler.press. Frankfurt a. M. (2015)
2. Gloger, B.: Scrum. Produkte zuverlässig und schnell entwickeln. Hanser Fachbuch, München (2008)
3. Godfrey, P.: Buliding a technical Leadership Model. 26th Annual INCOSE International Symposium (2016)

# Leading Teams to Top Performance

**19**

▶ In this chapter we describe step by step how to form high performance teams in product engineering. We start with the attitude of the leader and his objectives.

High performance teams are able to make remarkable contributions to the future of an organization. Leading a high performance team is a challenge for a conventional leader, because all credit for success goes to the team and not to the leader. If you can accept this, then you may be ready for the role of servant leader and facilitator. As a servant leader, you will become a service provider for the team. Ask yourself, "Do you want that?" If you answer "yes" to this question, be ready for something like a "sacrifice". High performance teams can only be formed with a servant leadership attitude.

## 19.1 Describe the Objective and Value Contribution of Your Team

In order to take a team of developers with you on your journey, the first step is to define a meaningful strategic objective. The goal must be attractive enough to inspire the team even before the start. Sketch it. Outline it. Precision is not necessary, perhaps even detrimental. The sketch of the strategic objective should generate emotions, and motivate the team. The team will learn more along the way and add additional detail to the objectives. This clarification should be made from the perspective of customers and stakeholders. What do they really want? What are their needs? We have therefore introduced the term "value contribution of the team". The value contribution deliberately reflects the ability your customer wants to achieve with your support and what it is worth to him (Fig. 5.2). Developers are usually oriented to create solutions. They think from the inside out, i.e.

© Springer-Verlag GmbH Germany, part of Springer Nature 2020
M. Jantzer et al., *The Art of Engineering Leadership*,
https://doi.org/10.1007/978-3-662-60384-0_19

from their abilities and ideas to the customer. There is nothing wrong with that approach. It is often successful, creating inspiring new solutions that customers could not even imagine. However, it often happens that good developers with good ideas do not open up a market. Sometimes competitors come up with unexpected solutions. Both maybe caused by thinking from the inside out.

By describing the value contribution, we encourage engineering leaders to think from the "outside in". What does the customer want to achieve? How can I support him?

Does the customer want a refrigerator? Or maybe he just wants fresh food anytime, champagne at just the right temperature, or cold beer. Are you already starting to invent alternatives to the classic refrigerator? This is exactly what we want to achieve: creativity, curiosity and a focus on the benefits for the customer. This is a necessary supplement to thinking from the inside out. It leads to a higher effectiveness in product engineering.

Once the desired value contribution to your customers is clear, reflect on what team you need to accomplish it. Teams need big challenges—a journey into the unknown— to reach the level of a high performing team. When everything is obvious, the team is focused on efficiency and lean processes. When everything is complicated or complex, the challenges are different. It is crucial to select the right team to meet the particular challenge at hand.

## 19.2   Composing Teams

Once you have clarified the objective and the value provided to the customer, you can start composing a team. The expression of targets and value contributions gives you a rough idea of the professional qualifications you need in the team. In addition, you should look for diversity.

Do you perhaps need additional employees from local markets who immediately understand what really is valued locally?

Are all functions and domains sufficiently considered: manufacturing process developers, purchasers, sales, product managers, designers, marketing, specialists? High performance requires a professional tension in the team, initial contradictions that have to be resolved. This is truly an advantage. The joint resolution of contradicting targets creates new and surprising solutions.

This step is about the people in the team and their personalities and is often difficult to grasp for engineering leaders.

To successfully assemble a team, you first need a good picture of each of your employees. What motivates them? What do they work for? What boundary conditions do they need in order to work optimally? To do this, leaders need to know a lot about their employees—and not only their professional qualifications, but their personal motivators.

Each member of a high-performance team has to join as an entire person and be accustomed to collaborating with different characters. As a leader, you should know how your potential team members interact socially. Have they already had positive experiences with each other or perhaps even bad ones? What do you know about their interactions before you became responsible? Have there been any private or professional conflicts between them? This could be an insurmountable obstacle, even if the competence profiles of the employees are excellent. The only way out is to change the composition of the team.

It is beneficial for the success of a team if it is diverse and balanced in its composition. Suggestions for team building can be found at Richard de Hoop [1]. He describes different characters in the form of musical instruments and advocates the conscious composition of an orchestra. Which strengths in interpersonal collaboration do you need in a team? The exact composition of the orchestra depends on the piece of music to be played.

Diversity in the team composition is often not considered sufficiently. Our experience teaches us that we can put top people in a team, and not much happens (see also [2]). When character strengths in a team are not properly balanced, a lot of energy is lost. We have met many managers who like to reject this: "Team members should behave professionally". One can wish for this, however "professional behavior" is often nothing more than avoidance of pain. This will not result in a high-performance team.

Balancing the characters of team members is an iterative process. People behave in groups with limited predictability. If it does not work as planned, talk to the team about it. Consider together what they can do differently. This may include exchanging individual team members. If you are transparent and empathetic, it is possible to do this without offending anyone. People want to be successful and understand that it is not about the individual but the composition of the team, its internal dynamics. For the development of the team and the individual it is important to talk about it.

---

**Example**

An example from our own team of leadership trainers: Our team initially consisted of, almost exclusively, very rational thinking persons, willing to perform. This was important and correct for the initial concept. Each of them could say exactly how to run a development organization. This led to controversial discussions, which did not always lead to shared results. Everyone could feel a certain dissatisfaction. To speak in the words of Richard de Hoop (see [1]): We already played different instruments, but for our piece of music we were not sufficiently diverse. In our self-reflection we came to the conclusion that we should strengthen around violins. We needed people who were more interested in the cohesion of a group, who cared about each other and who could bring in open and appreciating communication. We have specifically strengthened our team with violins. Today we deal with each other in a completely different way. We can all listen better, openly discuss issues, and appreciate diverse perspectives.

As a leader, you not only need to know a lot about technical content, but also about yourself and your employees. You should work to build an environment of trust, where people talk openly, and are comfortable revealing something about themselves. Only then do you have a chance to learn a lot about your employees. If you talk about what motivates and shapes you, then your employees will do the same.

## 19.3   Team Forming

Let us move on to another dimension: the time available. The members of a high-performance team have to adapt to each other and fight for the best solutions. That means they need time together. Team members of a high-performance team should be allocated 100% to the team. However, in our experience there is often a reluctance to assign the high performing experts 100% to the team, and they are allocated perhaps at only 20%. In this case it is better to organize the work differently. The experts can act as peer reviewers or coaches, for example. Alternatively, they can be "commissioned" by the team for individual tasks. There are many organizational alternatives to provide value while not being part of the team.

Why is full commitment and identification with the team and the common goal important? Because team members have to struggle with each other to find the "right" solution. They must be able to work out the seemingly contradicting requirements and fight for innovative solutions by disclosing and discussing conflicts. If they are part of a single team, they will most likely do so.

Product developers want to be successful, and that includes resolving contradictions. If they are in ten teams, they are unlikely to do so, because they simply have less time to resolve the conflict, and maybe they have one or two favorite projects where everything is running fine. They are already successful there. So why invest that much energy elsewhere?

---

**Example**

A few years ago, we had the dilemma that we had to start a new development of an innovative system quickly to hit a market entry window with the customer. The required experts were involved in other projects. Moving them would have endangered the other projects. Therefore, we started with a compromise and founded a "part-time team". Yes, there was progress, but it was much too slow.

We then restructured the development and put together a smaller 100% team of relatively inexperienced developers for the project. The team leader was the only one with real experience. In order to get the experience and the necessary knowledge into the team, we organized peer reviews with the experts. In the peer reviews, the development team had to explain what it had achieved in the current increment, its rough plan for the entire project, and the detailed plan for the next two to three increments. The experts' task was to ask questions about what had been achieved and to work out whether they considered the increments to be completed. On the other hand, they were only allowed to review the plan for completeness and ask questions about the proposed approaches. Their task was to help the team to improve the result and the plan, but not to assume responsibility. The responsibility remained with the team. During the review, the team had the opportunity to ask the experienced persons and experts for input.

This allowed us to vastly accelerate the development progress, even though we had few experienced developers in the team. After a few increments we had a powerful team. In addition, we had developed new forms of cooperation and learned how we become even more effective.

So do not hesitate to think outside the box and explore new working models.

## 19.4   Storming and Norming

Once the team is assembled and an initial working model has been defined, it is time to express your expectations on the team. What do you want from them? You have outlined the final result with the objectives and the value contribution. A good principle in requirements engineering is to define the acceptance criteria. This is also useful for specifying macroscopic development targets. This makes your targets more tangible for the team.

However, your expectations are not only about the result, but also about the way the team should work. Which methods should be used? How should increments and artifacts be documented? How should they involve you, your peers, your bosses? Which procedure do you want to see? What makes you happy? What would disappoint you? The demands on the team should be discussed with them, even those that seem obvious and self-evident, because they might be obvious only to you.

In this phase it is not only about your ideas, but also about the team, its boundary conditions and expectations. Even if the discussion does not bring much external progress, the time is well invested. The discussion of expectations and demands and the negotiation of mutual agreements reduces friction significantly and is an investment during the storming phase that will definitely pay off (Fig. 19.1). In our experience, leaders often lack the patience for such discussions. They want to see tangible results early, which is of course a good motivation. However, you win on the home stretch. For this you need a powerful start, which includes a strong focus on the storming and norming phase of team development.

Without this phase, there will be no high-performance team. Fighting for the best ways, aligning about solutions, and specifying targets are all linked to intense discussions. As a leader, you must promote this. In the end, everyone should "buy-in" to the goal and the approaches. These discussions are contentious, but must be allowed to happen.

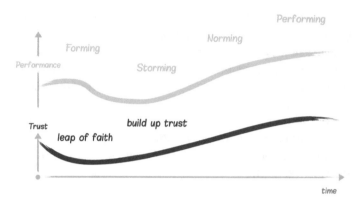

**Fig. 19.1**  Phases of team building [3]

At the same time, it is important to remain friendly and appreciative of each other. This significantly increases the probability for successfully forming a high-performance team.

---

**Example**

We were once asked to lead an interdisciplinary and cross-functional team to overcome a fundamental deficit in a product. It had to be fixed within nine months.

The team members were initially convinced that each had done everything right and that the problem had somehow been caused by the others. This was not helpful in finding a solution, but it was also a starting point to talk to each other. Each point of view was quite comprehensible and correct when taken in isolation, but each view was not complete. It represented only facets of the whole problem. For the first three months, we focused exclusively on developing a shared view of the whole problem and gradually figuring out who could contribute to the solution. After three months, we were able to decide without any remaining conflict who would do what to solve the problem.

We implemented the solution in the following six months, and of course, it didn't go perfectly. However, since we had transformed to a team based on mutual trust, this was no problem. Everyone on the team knew the other had done his best. After nine months, we turned a problem to a new market benchmark, which we enjoyed for 10 years.

---

Building trust in a team is crucial to success. Supporting the building of trust within the team is challenging enough, but also important is the team's credibility and trust with the stakeholders. Stakeholders have a right to be informed transparently. In our experience, this succeeds well if we divide the entire value proposition into increments and keep the promised value proposition per increment. If a change of direction is necessary, then this is explained to the stakeholders together with the new plan, and the new promise.

In order to make effective and efficient progress, decisions must be made where the data and findings arise (see Fig. 6.3). That means: The decisions have to be made in the team. The team thus assumes responsibility for the design.

At the same time, the team must take stakeholders with it, i.e. make transparent what it learns and what decisions it has made. Ultimately, the leaders are accountable for the results. To this end, the results must be available to the leaders. In addition, there will be decisions that the team can no longer carry alone or in which you as a leader want to be involved.

Find an agreement with the team on who decides what. What happens when the team reaches impediments, which they cannot solve on their own. Establish rules of procedure between the team and you. This creates clarity and commitment for all.

Once you have completed the phase, the team building process is complete. You have composed the team, accompanied and moderated the team-building phase and mutually agreed on a collaboration model. You have transformed a leap of faith into mutual trust. Now it is time to accompany the team to sustainable high performance.

## 19.5   The Performing Team

A team should take responsibility for the organization of its work as completely as possible. However, the accountability for results remains with the leaders. This responsibility is inseparably linked to leadership. Since you expect an excellent result, you have to place the design responsibility in the team. If you would permanently control the team, you would reduce their creative scope and slow down the design process. The team members would then have to invest time and effort in preparing reports for you. This is simply not cost effective. Leaders have the central task of building trust [4]. A trust that is not blind, because compliance must still be ensured, but a trust that motivates those to whom it is given. They strive to live up to it.

There are multiple ways to establish responsibility for design within the team and at the same time take accountability for the results.

Reviews are one strong way to achieve this. These reviews are not like examinations. Finding (supposed) errors or gaps is easy. Much more interesting are reviews that aim to improve the results for next increments. "Not to prove, but to improve". It is up to the team to decide if an idea improves the product. That is why the team has the last word in the review. Leaders do play a very active role in such reviews, sharing their experiences and perspectives, and making concrete suggestions - especially in areas where the team sees the need for improvement (see Chap. 14).

Another strong element is the retrospective. While the review deals with work results, the retrospective deals with the way of working. The aim of the retrospective is to become more effective and efficient. This requires practice and routine. Team-internal and cross-team conflicts must be uncovered. A good retrospective ends with a team agreement on the way of working for the next increment, or the team may identify impediments that it cannot solve. These are addressed to the responsible leader, who solves the issues. Leaders serve the team to win together.

The leader has to pay careful attention to what is handed over to him by the team. If there are truly impediments, which the team really cannot solve on its own, then it is your job. If there are problems that they could easily solve, challenge the team to solve them. Do not take on responsibilities that are better handled by the team. Usually, it is either insecurity or conflict avoidance. Both would be a topic for team coaching or even the retrospective.

If you want to see your team continue to grow, stay with it. It remains one of leadership's tasks to make the organization fit for the future. To do this, you need to know what is happening today and develop an idea of what may be possible tomorrow. This is necessary information and a consideration for strategy development.

In the Fig. 18.4 we have sketched a leader's journey to team empowerment. With a new team you start from scratch. At the beginning it is about learning what is important (black). Then working with the team follows. In this phase, leaders work almost as a team member. They contribute through decisions, experience and knowledge (magenta). Step by step the leader hands over responsibility. This is empowerment through coaching (blue). Regular reviews balance the responsibility for design and accountability for results between team and leader. Impediments are discussed and removed (green).

There are two behaviors that should be avoided:

- Micro-management means that you lack the confidence in your team to place enough decision-making power and responsibility with them. The team would be more efficient without you than with you. Some managers try to compensate for this by working even harder, but they remain the bottleneck. Giving trust and empowerment through coaching are big challenges for some leaders, especially since direct involvement provides a sense of personal effectiveness.
- The opposite risk is losing touch, which is caused by letting go too much. Even with completely empowered teams, leadership tasks remain. It is necessary to accompany the design process and to remove impediments on the way. A lack of attention over long periods can lead to the team becoming autonomous and unguided. Then they might lose overarching corporate objectives or miss stakeholder needs.

Fig. 19.1 once again graphically illustrates the process of team building. In our experience, people are always ready to start with a leap of faith in a new team. During the conflictual storming phase and the subsequent norming phase, this leap of faith is "transformed" into mutual trust. This development of trust is one of the keys to the team's high performance. The work of Lencioni [6], who studied cause-effect relationships for outstanding team performance, illustrates this. Figure 19.2 shows potential problems which keep teams from reaching their full potential and measures to overcome these

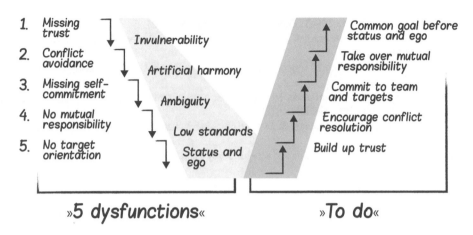

Fig. 19.2   The 5 dysfunctions of a team following [6]

problems. At this point, however, we would like to point out once again that tasks, people, team, company and country cultures are so diverse (see for example [5]) that mechanistic methods for forming high-performance teams will probably not work. Especially in social systems you will rather "probe—sense—respond" (compare—Fig. 16.3).

▶   **Practical Tips**
  • Maintain an open dialogue with your employees. First, build up trust, be approachable, and be fully present as a human being.
  • Ensure a transparent design process and review intermediate results. Empower the team step by step.
  • Measure team members primarily by team results.

---

**The Most Important in Brief**

Start with self-management. High performance teams need leaders who serve the team and its objectives. Leaders remove impediments that the team cannot remove.

High-performance teams need ambitious targets that they can and want to achieve together. A strong identity of the team must be ensured.

Pay attention to diversity of views and constructive conflict management in the team. Pay attention to a balance of characters and high social sensitivity

# References

1. de Hoop, Richard: Macht Musik. Gabal Verlag, Offenbach (2012)
2. Duhigg, C.: What Google learned from its Quest to Build the Perfect Team, New York Times Magazine, Feb 25, 2016
3. Tuckman, B.W.: Development sequence in small groups. Psychol. Bull. Berkeley **63**, 384–399 (1965)
4. Gräser, P.: Führung und Vertrauen. In: Keuper, F., Sommerlatte, T. (eds.) Vertrauesnsbasierte Führung. Springer Gabler, Heidelberg (2016)
5. Müthel, M.: Erfolgreiche Teamarbeit in deutsch-chinesischen Projekten. Springer, Heidelberg (2006)
6. Lencioni, P.: The Five Dysfunctions of a Team. Jossy-Bass, Hoboken, New Jersey (2002)

# Power, Influence and Win-Win

**20**

▶ To have power means being able to shape the world of others. We want to discuss the opportunities that come with power: leading your team, influencing your peer group, leading superiors, and leading stakeholders. Because conflicts do not fail to arise, we want to encourage leaders to tackle conflicts as early as possible to create real win-win situations. We present a proven method of negotiation, the Harvard concept, and apply it to handling technical contradictions.

The social psychologist Jonathan Haidt [1] has shown in his research that there are six common moral foundations across cultures. Authority is one of them. He explains this by the fact that authority facilitates cooperation within and between social groups, because decision-making is faster if it is clear who decides. Moreover, not everyone has a complete overview of a particular situation. Experienced people are trusted to make decisions in their domain and assume the associated risks. Hierarchy therefore forms naturally in social groups. As a rule, emergent hierarchies in groups are based on reliance, which in turn results from professional and personal aptitude. Leaders in a company are empowered from above. They enjoy the trust of the superior management. New leaders can bridge the resulting gap in acceptance at the team level, if they involve the informal leaders in the design responsibility.

In our experience, social influence happens above all through mutual inspiration, providing a sense of purpose, and building trust through integrity, and caring behavior. This works not only in the direction of one's team, but also on a collegial level (peer group) and upward through the hierarchy. Influential leaders are always involved in strong networks and act as respected interlocutors beyond their own area of responsibility (Fig. 20.1).

You should be well aware of your zone of influence and its limits. It typically covers more than the area of responsibility (Fig. 20.1). The zone of influence can and should be

© Springer-Verlag GmbH Germany, part of Springer Nature 2020    149
M. Jantzer et al., *The Art of Engineering Leadership*,
https://doi.org/10.1007/978-3-662-60384-0_20

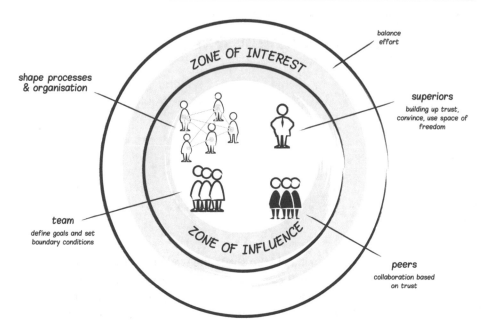

**Fig. 20.1** Team, stakeholders, superior hierarchy, own processes and organization are in zone of influence

used for the benefit of customers and other stakeholders. To invest effort outside one's zone of influence is usually not promising, but influence can be expanded gradually.

Influencing becomes challenging, perhaps even exciting, when leading intercultural teams or in a leadership role in a foreign culture. Power has a cultural dimension, that can be company—or country-specific. This dimension is called "power distance" [2]. High power distance means that power is unequally distributed. Small power distance shows a rather equal distribution of power. A discussion about the handling of power is highly recommended. The team should clarify the different expectations as precisely as possible and define the rules. Despite this approach, conflicts still occur. When they do, we recommend an approach based on a win-win strategy.

## 20.1  Win-Win

Conflict resolution and mediation seminars belong to standard leadership trainings. They primarily are about avoiding disputes. To put it simply: "Conflicts keep us from our actual work, therefore a leader must be capable of resolving conflicts". On the other side, conflicts in the form of technical contradictions are a source for creative solutions in engineering. With the right conflict resolution strategies, leaders can develop a relaxed attitude towards conflicts. This leads to effective cooperation and good technical solutions.

Conflicts are often a mixture of factual and personal conflicts. Although the principles on how to manage both types of conflicts are almost the same, a very first step is to separate the two.

Being in a conflict means emotional participation and effort. That is why people like to avoid them. However, very few conflicts resolve themselves—on the contrary. Friedrich Glasl [3] presents the temporal development of interpersonal conflicts as a downward staircase (Fig. 20.2).

Conflicts usually start because of different perspectives on a subject. In the early phase of conflict hardening, differences and contradictions become increasingly apparent. Points of view collide. This phase is necessary and helpful because it shows the potential for innovation and better satisfaction of stakeholders.

If opponents do not find a common solution, they will form opposing camps and increasingly polarize their views. At some point, black and white thinking dominates the argumentation. Instead of an informal exchange of ideas, the opponent is cornered during debates.

As soon as the feeling prevails that debates and talks are no longer fruitful, the parties act unilaterally and create facts. The opponent is intimidated and offended. In this phase of conflict, win-win solutions are still possible, i.e. both parties in conflict can still emerge as winners, but the leader must act now.

Otherwise, a battle for image and credibility begins. The opponents drive themselves by insinuations into a loss of face. The opposing parties seek to stage conflicts in

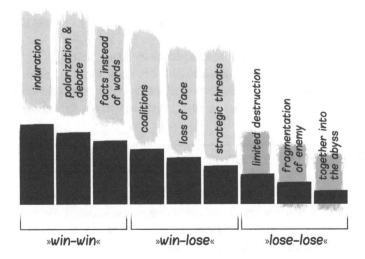

Fig. 20.2  Escalation of conflicts according Glasl [3]

public to gain support but may actually drive outsiders away. The loss of trust becomes complete and reaches a point of no return. The potential for violence increases through undisguised threats and counter-threats. The parties involved become increasingly radicalized. From this phase not everyone can emerge victorious, the damage already suffered is too great. Either the involved parties must be permanently separated, or one of them has to be let go.

If this does not happen, the opponents continue to harm themselves. Finally, they may even accept damaging themselves, as long as the damage of the opponent is greater. In the fragmentation phase the opponents try to annihilate each other. In the end the conflicting parties are even prepared to accept their own annihilation just to defeat the opponent. In this phase there are only losers. This terminal conflict phase can only be ended by massive external intervention. Such conflicts are a waste of company resources.

In the professional environment, things must not get this far. Leaders must intervene, exercise their formal power and refocus the work on the customer.

The sooner a conflict is resolved, the greater the chance for a win-win solution. First, it makes sense to clarify the type of conflict and to take the first step towards a solution:

- Role conflicts arise in the event of unclear responsibilities, e.g. line or project.
- A classic conflict of objectives exists if goals are mutually exclusive, for example, high in-house production share vs. a high proportion of third-party components.
- If one party wants to solve a problem experimentally and the other by means of simulation, there is a procedural conflict.
- A conflict of interest exists if the interests behind the object of conflict are contradictory, e.g. "a positive annual report" vs. "dealing openly with failures".
- Conflicts of values are about what is "important" to the parties, e.g. sustainability vs. rapid success?
- Disputes over scarce resources (budget, resources, laboratory facilities) lead to distribution conflicts.
- Relationship conflicts can arise if role expectations do not match.
- Conflicts can also occur within a person. What is more important to me as a person: "supporting the team and working overtime" or "spending the evening with the family"? We call this an intrapersonal conflict, which we will not discuss further here.

## 20.2  The Harvard Concept

The Harvard concept of negotiation [4] is considered the most effective negotiation technique. The aim is a win-win result and not a classic compromise. Both sides strive to achieve the greatest possible benefit. Beyond an agreement on merits, a lasting improvement in the quality of a relationship can be achieved.

The Harvard method consistently identifies two levels of communication, the object of negotiation and the relationship, and is characterized by four core steps:

1. People and their interests are taken seriously. Negotiating "hard" is allowed but must focus on the issues. The interpersonal contact remains respectful and free of injury. It is based on understanding, trust and the ability to put oneself in the position of others and to integrate one's perspective.
2. The interests of the parties involved are in focus and not their points of view. Right and wrong are not the primary concern; instead, the true interests of the parties are in focus. These can be goals, wishes, constraints and worries. This negotiating step is characterized by questions. The disclosure of the own interests is an important door opener.
3. As many decision options as possible are collected and developed, e.g. in a joint brainstorming session, offering a wide range of options.
4. After agreement on common objective assessment criteria (e.g. legal regulations or judgements, ethical standards, market value, comprehensible costs), the decision options can be evaluated and selected. This step increases the acceptance of the selected solution.

Even with the best intentions, there is always a risk that the negotiation will end in a poor agreement. Fisher and Ury recommend finding the "best alternative to a negotiated agreement" (BATNA). It can be used repeatedly during the negotiation for comparison with the current agreement.

---

**Example**

During a central platform development, we were completely overloaded by the number of issues to be solved. Each department had so many problems to solve that everyone had started looking into only their own issues. More just wasn't possible. Focusing on the own topics was important and understandable. Interdisciplinary cooperation nearly came to a standstill. The willingness to support a colleague slowly approached zero. As a consequence, the success of the overall project was massively endangered.

What could we do? Cancel the project, reduce the goals, change the teams? These were all options, but with undesirable consequences. The division's management decided to break the deadlock by applying the Harvard concept.

The first step was to get the growing interpersonal conflicts out of the way, to recognize the colleagues not as resources for additional tasks but as human beings. We invested time in joint evenings and events that simply put people first and not their role in the company. Since we all knew each other, this was rather easy.

Next, everyone individually collected his or her biggest challenges. We explained to each other what we really needed to solve our problems. Our management gave us guardrails within which we had to find a way forward: additional budget and resources were excluded, and customer schedules had to be maintained. New priorities within the project objectives were permissible, if they could be reasonably justified.

We now had a focus on the technical and organizational problem. We then discussed potential solutions, and more importantly we talked about the overall project objectives and priorities. Thus, we had again found a basis to discuss the whole problem from a higher point of view.

The customer-relevant goals of the platform project were fully achieved by the start of series production, and we achieved our internal goals two years later—according to the agreed priorities. The product family is still running very successfully in the market today.

## 20.3   Dealing with Technical Contradictions

We now apply the principles of the Harvard Concept of changing perspectives and looking at deeper interests to situations where we are confronted with technical contradictions. We take the task of converting direct current from a battery into alternating current for the motor of an electric vehicle as an example. High electrical output power requires high currents. Heat loss is generated during conversion, which in turn reduces efficiency. This can be compensated by cooling measures, but they are expensive. Many domains are involved in the solution of this engineering task, e.g. circuit designers, developers of power components, mechanical designers, developers of the manufacturing process. Each of them first looks at the problem from their own point of view.

Using the functional view (Fig. 20.3) you can search for a solution in a domain-neutral manner (corresponds approximately to step 1 of the Harvard concept). The underlying technical interest is found in the functions of the superordinate system level: "switching power" and "limit temperature".

On the component level, the solution is provided by thermally conductive material and an efficient coolant flow. The level above requires a limitation of the maximum temperature of the power modules, which may only be reached during a few driving maneuvers.

Additional options arise from even higher system levels. For example, voltage peaks and thus power peaks can be reduced by the way the electronic switches are controlled. The peak generated heat is thus reduced, thereby eliminating the need for special control functions or additional dissipation measures.

The solution space suddenly expands for creative and innovative approaches (step 3). These can be evaluated in terms of customer requirements, unique selling points, costs and/or risks (step 4). A jointly developed and sustained decision is a good basis for the cross-domain development of the solution.

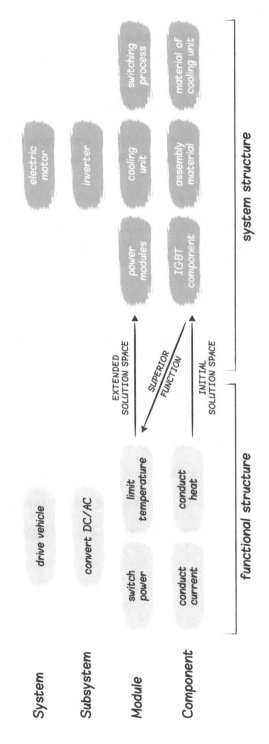

**Fig. 20.3** Finding solutions with the help of the system and function view

▶   **Practical Tips**
  • Consider how you can increase your zone of influence: who else is impor-
    tant outside your team? Create a stakeholder map (Fig. 8.2).
  • Observe your own behavior and that of your team with regard to conflicts.
    Be passionate and appreciative about win-win solutions.
  • In conflicts, explore the underlying interests of all parties and look for solu-
    tions or at least improvements.
  • Apply the Harvard principle to your technical contradictions: what are the
    interests in the overall system (functional requirements)? Are there alterna-
    tive solution options?

---

**The Most Important in Brief**

To have power means being able to shape the world of others. In our daily work, power stands for influence: influencing the peer group, leading superiors, leading stakeholders. Conflicts are inevitable. The earlier these are addressed, the wider the solution space and the better the chance of a win-win result.

The Harvard concept is a useful concept for conflict management. Even partial steps can help. When leaders develop a constructive attitude toward conflicts, they create a working environment that is capable of dealing with technical contradictions. This enables technically superior solutions in the end.

---

## References

1. Haidt, J.: The new synthesis in moral psychology. Science **316**, 998–1002 (2007). https://doi.org/10.1126/science.1137651
2. Hofstede, G.: Culture's Consequences. Sage Library of Social Research, Beverly Hills, CA (1980)
3. Glasl, F.: Konfliktmanagement. Diagnose und Behandlung von Konflikten in Organisationen. Haupt, Bern (1980)
4. Fisher, R., Ury, W., Patton, B.: Das Harvard-Konzept. Der Klassiker der Verhandlungstechnik. Campus, Frankfurt a. M. (2013)

# Developing Leadership Excellence

# 21

▶ We describe some of the aspects of "leading myself" that are important to us. They are the basis for leading others. Effective self-leadership is based on knowledge of one's values, strengths and individual way of learning.

We present a tool, the "leadership model", to develop excellence in leadership. The leadership model supports the discussions about leadership in an organization. It fosters the formulation of common expectations, the definition of procedures and the prioritization of tasks.

## 21.1 Value, Strengths, Ability to Learn

Leaders need to know themselves well to be able to lead others consciously and responsibly [1, 2]:

- Their own framework of values serves as an inner compass for decisions.
- They need to know and leverage their own strengths and talents. They can compensate for their own weaknesses through employees with complementary skills.
- Developing and leading others requires a learning process for the leader. Leaders who know their own style of learning are more efficient.

**What are My Values?** Each one of us has a set of values, which forms and changes over the course of our life. Our social environment, family, society, and so on, all shape our view of what is important.

Values influence all decisions. Leaders operate in a framework that is provided by the values of the company. If personal values do not match those of the organization, internal

© Springer-Verlag GmbH Germany, part of Springer Nature 2020
M. Jantzer et al., *The Art of Engineering Leadership*,
https://doi.org/10.1007/978-3-662-60384-0_21

conflicts are inevitable, and these impede performance. In addition, there is a risk of losing credibility ("act with integrity").

A harmonious framework of values is a prerequisite for excellent work results.

---

**Example**

As a company in an agile transformation, we at Bosch are currently experiencing a change of values: among other things, *(planning) stability* and *welcoming change* are being reevaluated.

Every leader has to decide how to deal with this change in values. In discussions with leaders, there is often considerable resistance to the new values. It's no wonder, since *planning stability* was highly valued for many years. As an example, some time ago the planning horizon for the annual budget was reduced from three years to one year. This was an important signal that values were changing.

Some are not prepared to adjust their value systems and leave the company in such transitional phases. Others waste a lot of energy complaining about it. As a leader, this is detrimental. If you pay lip service to the new values, you lose credibility with your team. If you reject them, you lose credibility with those propagating the new values. How should the employees be convinced to *welcome change* when you do not?

Those who consciously deal with the new values and find their own position also orient their team quickly. They enable the team to act in the new value system. You can define for which work products *planning stability* is still important and in which situations flexible reaction to changes is required.

**What are My Strengths?** You probably know your weaknesses and have already invested a lot of energy in changing or improving them. Surely, you have wrestled with a few. One should definitely know one's weaknesses, because they will not contribute to success. However, if you know them, you can develop strategies to mitigate their impact e.g. by ensuring that your immediate team has employees with complementary skills.

Leaders usually deploy their employees effectively and make use of their strengths. In our experience, however, they rarely do it for themselves. From a purely economic point of view, however, this is a mistake. The further development of strengths is far more efficient than working on weaknesses. It can lead to achieving an excellent work result with a more significant impact than an incremental improvement in a weakness. To maximize your personal effectiveness, it always makes sense to get feedback on your strengths (e.g. [3]). If you know your strengths well enough, the next question is where you can use them best.

**What is My Way of Learning?** Learning means receiving information and processing it in such a way that we can complete a task or solve a problem. There are many different learning strategies such as reading or listening. We process information, for example, in the form of pictures or as a sound sequence. Some learn intellectually, others rather physically, or through experience. The environment is also a factor, as some learn best when alone, and others learn best in a group.

Reflect on your own learning style. Watch how your employees can learn best and design an environment that will give both you and your team the best possible environment to learn.

## 21.2   The Leadership Model

In this book we have collected what we consider the most important aspects of leadership in engineering. As if from a menu, you can select what is important to you. We have tried to supplement the theory with examples from our consulting practice. These should be adapted to your specific leadership tasks and the environment of your organization. For this purpose, we have developed a leadership model that is particularly suitable for leaders with design responsibility for technical solutions (Fig. 21.1).

The starting point is the strategic objectives of the business unit, which are derived from the overall corporate strategy. The value contributions of the business division and the leader are derived from these. The aim is to achieve excellent results, using various levers and procedures as well as skills and competences.

In our trainings, we use the leadership model which has its origin in the EFQM excellence model [4] as a tool to reflect leadership in one's area of responsibility. The leadership model shows (like the EFQM model) a results part (right) and an enabler part (left):

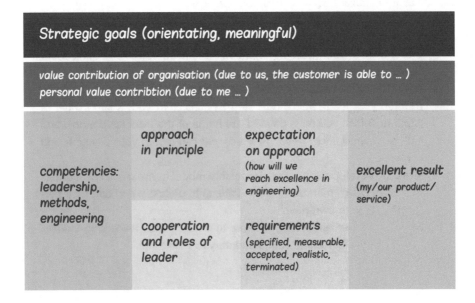

Fig. 21.1   A leadership model for excellence

Products and services that a leader and his team develop are presented in the results section (far right), including the criteria for excellence (middle right). We have also introduced a field that describes the expectations on the way products are developed. In our view, this is a key lever for achieving excellent results in product engineering.

"Enablers" are technical, methodical and leadership skills (left) as well as a systematic approach in product engineering. The interaction within the organization and the expression of a suitable role understanding (center left) contribute to the excellent result.

In our seminars, the leadership models are developed over the course of approximately one year. For most of our seminar participants, it is a new experience to show explicitly how they see themselves in their leadership role. Clarifications with other parties involved in product engineering are particularly helpful. During this phase, several revisions take place. We often observe reprioritizing of operational tasks when leaders of the same organization focus on concerted goals and value contributions.

On a personal level, a common feedback from participants is "my role is now much clearer to me" or "I can now act more effectively", or "I have significantly expanded my zone of influence". Some leaders also use it to make their team's own standards clear. Teams also appreciate this: "now we understand much better where our leader wants to go with us and why".

The leadership model can help to lead more effectively on an individual level. It helps in determining one's point of view, in working out expectations, and in prioritizing procedures.

Once leaders have learned to discuss their leadership models with each other and their employees, they can more easily adapt their leadership style, their engineering framework, and their organizational units to new challenges.

▶   **Practical Tips**
- Create a list of your most important values and strengths and think about how best to use them.
- Adapt the communication within your team to the different learning styles.
- Develop your leadership model step by step. Start with a large sheet of paper on which you have painted the frame of the leadership model and a block of small sticky notes on which you can record your thoughts and impulses.
- Make the development of your leadership model transparent. Invite others (superiors, peers, mentors, team members) to discuss it and add their perspectives to your own view.
- Encourage other leaders in your area to develop their leadership models, making them aligned and complementary to each other.

> **The Most Important in Brief**
>
> Leadership of others is based on leading oneself. By knowing their personal values, strengths and very individual way of learning, leaders can act consciously and effectively.
>
> The leadership model is a tool for developing leadership excellence in product engineering. It is created individually and used to communicate leadership specifics within a company. It helps to lead more effectively, to describe a common expectation and to bundle and prioritize approaches.
>
> By discussing their leadership models, leaders can adapt to changes from outside and inside in a flexible and sustainable way.

# References

1. Drucker P.F.: Managing Oneself: Harvard Business Review, Jan. 2005, p. 2
2. Corssen, J.: Der Selbst-Entwickler: Das Corssen Seminar. Verlagshaus Römerweg, Wiesbaden (2004)
3. De Hoop, R.: Macht Musik. Gabal, Offenbach (2012)
4. Die Grundkonzepte der Exzellenz. EFQM, ISBN 90-5236-079-0; Brüssel (2003)

# Index

© Springer-Verlag GmbH Germany, part of Springer Nature 2020
M. Jantzer et al., *The Art of Engineering Leadership,*
https://doi.org/10.1007/978-3-662-60384-0

# Ihr kostenloses eBook

Vielen Dank für den Kauf dieses Buches. Sie haben die Möglichkeit, das eBook zu diesem Titel kostenlos zu nutzen. Das eBook können Sie dauerhaft in Ihrem persönlichen, digitalen Bücherregal auf **springer.com** speichern, oder es auf Ihren PC/Tablet/eReader herunterladen.

1. Gehen Sie auf **www.springer.com** und loggen Sie sich ein. Falls Sie noch kein Kundenkonto haben, registrieren Sie sich bitte auf der Webseite.
2. Geben Sie die eISBN (siehe unten) in das Suchfeld ein und klicken Sie auf den angezeigten Titel. Legen Sie im nächsten Schritt das eBook über **eBook kaufen** in Ihren Warenkorb. Klicken Sie auf **Warenkorb und zur Kasse gehen**.
3. Geben Sie in das Feld **Coupon/Token** Ihren persönlichen Coupon ein, den Sie unten auf dieser Seite finden. Der Coupon wird vom System erkannt und der Preis auf 0,00 Euro reduziert.
4. Klicken Sie auf **Weiter zur Anmeldung**. Geben Sie Ihre Adressdaten ein und klicken Sie auf **Details speichern und fortfahren**.
5. Klicken Sie nun auf **kostenfrei bestellen**.
6. Sie können das eBook nun auf der Bestätigungsseite herunterladen und auf einem Gerät Ihrer Wahl lesen. Das eBook bleibt dauerhaft in Ihrem digitalen Bücherregal gespeichert. Zudem können Sie das eBook zu jedem späteren Zeitpunkt über Ihr Bücherregal herunterladen. Das Bücherregal erreichen Sie, wenn Sie im oberen Teil der Webseite auf Ihren Namen klicken und dort **Mein Bücherregal** auswählen.

## EBOOK INSIDE

**eISBN**
**Ihr persönlicher Coupon**

Sollte der Coupon fehlen oder nicht funktionieren, senden Sie uns bitte eine E-Mail mit dem Betreff: **eBook inside** an **customerservice@springer.com**.

Printed in the United States
by Baker & Taylor Publisher Services